"*Connections* takes its place on my short list of intelligent, thoughtful and eminently sensible books about alien abductions. Collings and Jamerson present their decades-long, mutually interlocking UFO experiences in a dramatic, highly readable and well-illustrated account. Despite the disturbing nature of their encounters, it is the authors' humanity and liberating common sense that finally affect us so deeply."

> Budd Hopkins
> Author of *Intruders* and *Missing Time*

"[*Connections*] is a first-rate piece of work that I'd be proud to sign my name to."

> Don Berliner
> UFO Investigator

"Beth Collings and Anna Jamerson tell their stories with great courage and clarity. Here is a wealth of material we can all use when looking for explanations of the great mystery of our times—alien abductions. May we all join Collings and Jamerson in seeking 'Connections.'"

> Andrea Pritchard
> Principal Editor of *Alien Discussions: Proceedings of the Abduction Study Conference held at MIT*

"An utterly captivating and articulate, intimate look into the thoughts and emotions of two women who personally share their experiences and reactions when confronted with the UFO abduction phenomenon. A sure catharsis for abductees and a treasure-trove of data for researchers."

> Raymond E. Fowler
> Author of *The Andreasson Affair, Phase I and II, The Allagash Abductions, The Watchers* and *The Watchers II*

Connections

Solving Our Alien Abduction Mystery

by Beth Collings
and Anna Jamerson

Wild Flower Press
P. O. Box 726
Newberg, OR 97132

Connections: solving our alien abduction mystery / by Anna Jamerson and Beth Collings.
p. cm.

ISBN 0-926524-35-6: (alk. paper) $17.95

1. Unidentified flying objects--Sightings and encounters. 2. Life on other planets. 3. Abduction. 4. Jamerson, Anna, 1949-. 5. Collings, Beth, 1946-. I. Collings, Beth, 1946-. II. Title.

TL789.3.J36 1996
001.9'42--dc20

95-49150
CIP

DESIGN IMPLEMENTATION: Carlene Lynch
COVER ART & DRAWINGS: Beth Collings
COVER DESIGN: Maynard Demmon

Printed in the United States of America.

Address all inquiries:
Wild Flower Press
P. O. Box 726
Newberg, OR 97132
U.S.A.

Printed on recycled paper.

Dedication

To the Fund for UFO Research,
with special thanks to Richard Hall,
Rob Swiatek, Don Berliner and friends.

Dedication

To the Fund for UFO Research,
with special thanks to Richard Hall,
Rob Swiatek, Don Berliner and friends.

Table of Contents

Introduction

Early in 1992 when Anna found us listed in the Washington, D.C. area phone book under "UFO-Fund for Research," I had been deeply involved for the preceding four years in conducting a pilot study of alien abduction reports for the Fund. Rob Swiatek, another member of the Fund Executive Committee, and I had interviewed more than forty abductees from around the Washington, D.C. area.

We had already experienced a "data overload" and had gone through a period of burnout from dealing with the emotional turmoil engendered by the abduction experience. As reflected in this book, the abductees wanted answers (meaning ultimate explanations of what was going on and what it all meant), and, of course, we didn't have any. Their needs were far greater than anything we could supply.

It had reached a point where another phone call from another abductee was not exactly greeted with enthusiasm. However, since we were determined to be professional in our study, Anna and Beth were about to become our latest "statistics." No one was to be turned away. Don Berliner, who mans the Fund office phone (the office being the corner of his apartment in Alexandria, Virginia) called to brief me about his conversation with Anna. As it happened, I had a bad cold and hadn't felt up to going on an interview, so I urged Don to go along with Rob. They drove out to Anna's horse farm and conducted the initial interview—the first of dozens of interviews and meetings with Beth and Anna that followed over the next several years.

When Rob and Don later briefed me about the interview, they told me the story of Beth's Christmas season encounter with all its bizarre features. Despite the content of the story, they had been favorably impressed by Anna and Beth as far as their sincerity and apparent credibility were concerned. Having listened to similar reports for several years, we were no longer fazed by the weirdness of story content—that was a standard feature.

The pilot study was designed to learn the parameters of the abduction experience and to develop some guidelines on how to go about studying it. Our approach was to try to be good listeners and record everything the abductees told us, while offering whatever help, advice and moral support we could. Typically their experiences would be ongoing, so we would remain in touch with them over a long period of time. We had quickly learned that standard scientific investigation techniques did not always apply well to abduction reports, and that the experiencers usually were frightened, confused and disoriented. They needed psychological support, not intrusive and uncaring investigative probing into their lives.

Our network included a few psychotherapists, but far too few for the numbers of experiencers we were working with. Many abductees could not afford the professional fees, or did not trust psychologists who often treated them as if their reported experiences were fantasies and they were "sick."

Despite the importance of the support role, we were determined not to lose our critical faculties, to simply accept as truth anything we were told. So we resolved to provide a sympathetic ear while getting to know the experiencers over the long term and evaluating their credibility and "sanity" (for lack of a better description). How did they react to and deal with abduction experiences? Were there any overt signs of individual or family group pathology? Was there any indication of ulterior motives? Was there any tangible evidence to support what they claimed?

A few of the people we dealt with strongly needed professional psychological counseling (which did not prove one way or the other the reality/unreality of their reported experiences). Some could not deal with what was happening to them at all, and dropped out. Most, however, were amazingly resilient and—though continuing to experience frightening and disturbing events—found ways to cope with them and to function in their normal day-to-day lives.

By the time of the Abduction Conference at the Massachusetts Institute of Technology (M.I.T.) in June 1992 (the proceedings of which have been published and are now available[1]), I had become so favorably impressed by Anna and Beth that—as a member of the Conference Committee—I prevailed on the conference committee chairmen David E. Prichard and Dr. John E. Mack to invite them to the conference. I did so because Beth and Anna were so intelligent and articulate, in my estimation, that I thought they would make outstanding witnesses to the reality of the abduction phenomenon before the scholarly audience at M.I.T.

By this time, there was no doubt in my mind that the abduction phenomenon was "real," and that if we were going to try to explain

---

1. Pritchard, Andrea, ed., et al. *Alien Discussions: Proceedings of the Abduction Study Conference.* Cambridge, MA: North Cambridge Press, 1994.

it, their testimony was important. Their verbal and artistic skills enabled them to express clearly and convincingly what it was like to be an abductee. Of the dozen or more abductees who had been invited to participate in the conference, Anna and Beth were the only ones who presented formal papers. They also participated in a panel discussion with the other abductees.

What has impressed me all along about Beth and Anna is the very healthy self-doubt they have even as they recount their experiences. Was that real? Did I only dream it? Am I being influenced by other stories that I have heard? This reflective attitude is evident in their writings and is a credit to them, intellectually. Clearly, false elements do creep into memories and not everything that abductees *think* is real can be taken at face value. An important part of the research problem is to separate apparent facts from false memories.

Because of the unique nature of the phenomenon, abductees become co-investigators with UFO researchers and therapists, as is the case with Anna and Beth. The roles of researcher and therapist also tend to overlap. The triad of experiencer, investigator and therapist working together has proven to be an effective way of studying the abduction phenomenon while providing the support necessary for dealing with its personal and emotional impact on the individual and the family.

In the course of working together with Beth and Anna for several years, we have become friends as well as colleagues in the search for answers. All of our interactions with them have indicated their sincerity, honesty and truthfulness. They are as puzzled about the ultimate meaning of it all as we are.

Whatever the true nature of the abduction experience ultimately proves to be, I commend this beautifully written and candid book as an honest attempt by two very admirable people to convey to the world at large the nature of the abduction experience and its powerful impact on people's lives. It takes courage to learn to live with the intrusive and discomfiting procedures that abductees are forced to undergo, and to resolutely seek answers, even in the face of widespread skepticism. Anna and Beth's courage shines through the pages of this remarkable book.

Richard Hall
Fund for UFO Research, Washington, D.C.

Preface

This book has been written in first person throughout because it is our story throughout. Our purpose is to describe our personal feelings, experiences, thoughts and reactions to these events in a sincere and candid fashion; our hope is to offer solace, empathy and encouragement to other experiencers who need to know they are not alone.

We are both women in our mid-forties who believed until recently that we had led lives no more—or less—remarkable than most. What is remarkable is our childhood connection associated with our earliest memories of alien abductions. When we first met as adults, neither of us realized we had known each other since we were little more than toddlers. These memories of early ties did not surface until many years later, after intense alien activity provided the catalyst. Had our meeting in later life been prearranged, our destinies intertwined for some purpose? We don't know yet. We do know circumstances leading to our second meeting stretch coincidence to the extreme.

How this relates to the alien abduction phenomenon we can't—and won't—speculate. We will, however, suggest it is not chance, although its ultimate purpose eludes us.

We do not propose there is but one explanation for what is happening to us and to so many others out there. We do not presume to understand it, nor do we have a clue how to make it stop. We do, however, have a strong will to survive, and it is this we wish to impress upon the reader. We will continue to resist this intrusion into our daily lives and the lives of our families in the hope some future event will make all things clear.

It would be far simpler, and less stressful, to leave well enough alone, to write these strange alien encounters off to imaginings, fantasy and illusion. But we both feel this would create more problems than it would solve. Facing reality has taken on a whole new meaning for us now; this other reality has forced us to evolve, to adapt.

It was vital that we learned to deal with the emotional backwash of these experiences. Fear feeds on itself, and fear of the unknown closes all doors to understanding. If we were ever to come to grips with this phenomenon, we would have to accept our experiences—

good or bad—as a part of life. This acceptance led to healing: We forgave ourselves for being abductees!

The change in perspective allowed us to view our world and ourselves differently. We realized that some of our abduction experiences—even the most upsetting—have benefited us in ways we could not have imagined a few years ago. Our perceptions are more acute, our minds more open. Things we once took for granted have become priceless; problems we once thought insurmountable seem less significant. We have discovered not only the strength that comes from within, but also the strength that comes from our most valuable resource: Friendship.

Thanks to the support of our families and friends, the encouragement and understanding of investigators, therapists and hypnotists, we have gradually begun to put our lives into perspective. We do not expect concrete answers to this enigma, but in our search for truth and understanding we have discovered ourselves.

We are no longer who we were. We will never be again.

Beth Collings and Anna Jamerson

Prologue

Beth

This is the most difficult thing I have ever done, putting down in words what even I myself find fantastic, if not bordering on the bizarre. It is made doubly onerous by the subject's very nature: alien abductions. This is a topic more likely to be bantered about by children trying to frighten themselves with ghost stories and tales of witches and ghouls than by a mature woman who has (until recent times) had a firm grip on reality and all it entails. There was no question of who I was, how I came to be, or that my destiny was in my hands alone. I was to discover these "knowns" were not necessarily static—or real.

What has brought about this reshaping of my reality? Aliens.

Incredible? Unbelievable? Indeed.

Nothing short of solid, irrefutable proof could have convinced me that such things existed in our realm. The subject of UFOs was no more than a euphemism, a way for the human mind to categorize something with which it could not relate or identify. The subject, fascinating as it might be, held little interest for me—before.

I saw the movie *E.T. The Extraterrestrial.* I stood in line in the rain for over an hour to see *Close Encounters of the Third Kind* just to verify that the special effects were as good as the critics claimed. I enjoyed both of these movies, have even watched them again on television. But that's fantasy. That's the movies.

I've always entertained the belief that we cannot be alone in this vast universe. The human race is but a mere speck of sand on an endless beach, so to assume our little planet contains the universe's only sentient species is utterly ridiculous. It is certainly plausible that UFOs exist (and many people may have seen them), so perhaps in the not-so-distant future we may encounter otherworldly species in our continuing exploration of our planetary system—and beyond.

But I stopped short of considering any possibility that alien beings might have come to us first. Ha! What hogwash! No one in his or her right mind would believe that there were real ETs in the back yard scrambling for Reese's Pieces. And, of course, none are.

These aliens scramble for no one. They have their own agenda and will abide by it whether or not we humans cooperate. They do not offer friendship, compassion or loyalty; they are not cute and helpless. They do not follow our expectations of how aliens should look or behave. They are alien in every sense of the word.

It is this frightening reality that has turned my sane world into one of total chaos and uncertainty. What is real? What is fantasy? What is *crazy*? How can one tell the difference when faced with such unearthly creatures? Witnessing a UFO zip over your house or buzz your car becomes a fascinating story to tell friends and family, an experience that when related in the company of total strangers is now considered socially acceptable. These days, describing unidentified lights in the night sky (or even possessing video tapes *proving* what you claim to have seen) generates little more than bored yawns from once captivated listeners who have heard it all before, and probably seen it before, too. A series of television programs have been dedicated to this subject for a number of years; commercials using UFO and alien themes abound; magazines and newspapers report sightings, encounters and even bizarre tales of alien intervention. If one has access to any media these days, the odds of remaining ignorant on the subject (whether one is a believer or not) are extremely slim.

But what happens when one relates a story about encounters with small gray slender beings with oversized heads, huge black eyes and no toes who float their paralyzed human victims through closed bedroom windows in the wee hours of morning? I'll tell you what happens: The teller is unceremoniously categorized as *crazy as a rabid hound*, a person to be *shunned*, a person in dire need of *professional counseling*, even a person who is possibly a victim of childhood abuse.

The abductee (or experiencer) is given few options when faced with these presumptions. My first reaction was to deny anything unusual had ever occurred. (I return to this protective mechanism on a regular basis.) From that point I moved on to exploring other possible explanations for the phenomenon, more earthly interpretations that included lucid dreams, exhaustion, stress, eye strain, flights of fancy, low self-esteem, paranoia, etc. The list went on and on, but eventually these had to be abandoned in favor of a less popular theory: mental illness. Though not desired, this explanation was preferable to alien intervention. Besides, mental illness was treatable with a good chance of complete recovery. I could expect to return to a normal, alien-free existence, put all this nonsense behind me and go back to the process of living. I could even discuss my illness and treatment in public since mental illness was far more acceptable. This would have been the final chapter had the therapists found any pathology, anything at all that might explain these events in psychological terms. But they found none.

Well, there was always frontal lobe disorder or some other physiological malfunction causing these weird visions. This would be my next avenue of investigation. After numerous—and costly— physical exams by qualified (and baffled) physicians, I was told there was nothing wrong with me, but perhaps I should get more rest. This, I was certain, was their way of saying, "We can't treat psychoses. You need to see a shrink."

With these alternatives gone, I was faced with the one explanation that I just couldn't bring myself to accept: I was being routinely abducted by aliens, had been since early childhood, and there was not one thing I could do about it. I could not dial 911 for help; I could not take a pill and hope it would all go away; I could not ignore the fact that something truly mysterious was going on. In order to deal with it, I would have to change my whole way of interpreting reality.

Even so, my mind refused to surrender to the idea of an alternate reality. In defiance, I declared it was a case of vivid imagination. I would simply deny the existence of little gray aliens who went around abducting humans for experimentation—no matter what strange memories plagued me. I would force myself to forget, command my memories of unexplained and frightening events to cease and desist, and generally *get a life*.

But human curiosity does not fade away on command, and my nature was of the most inquisitive kind. I come from a family of artists and explorers, a people who have for generations been unable to resist challenging the unknown, learning all that can be learned, striving for greater knowledge, no matter what the sacrifice. I have followed in their footsteps; my grown son has followed in mine; my granddaughter has taken her first steps on that same path. Who was I to think I could circumvent the inevitable? I needed to know what was happening. I needed to believe in one thing or the other. These experiences were either real (strange as they may be) or they were imaginary. They could not be both.

Armed with this new determination, I began documenting everything. Each and every unexplained wound, bruise, burn and scar was photographed and cataloged. I confided in my parents and took notes on their experiences, questioned them relentlessly in an effort to clarify many disturbing childhood memories. I discovered soon after (and to my great dismay) that my granddaughter, Noel,[1] had

---

1. The names of all family members and some friends have been changed throughout the course of this book.

been enduring her own alien-related episodes since the tender age of three.

The more I learned, the less I felt I could handle. Things were far more complex than I had at first surmised—and far more widespread. How many others out there were going through the same types of experiences?

Thousands, perhaps millions, I learned. I was not alone after all. This understanding, if nothing else, prevented me from collapsing under the strain of continued interference by the aliens. I planned to survive all this. I would survive it by documenting, by striving to assimilate, by confiding openly with friends and family, and by taking advantage of my own innate tenacity.

The aliens may be able to maintain complete control while I'm in their domain, but in mine I will take charge!

Anna

Tired of the same old thing day after day? In a rut? Looking for some excitement in your life? Being abducted by aliens will surely change your life, but almost any alternative is preferable. I long for the sameness and boredom I used to detest. Excitement, fascination, depression and terror seem to overwhelm me almost daily as I begin to accept a new reality that the majority of the population has little reason to believe exists. I never believed anything like this existed outside the covers of science fiction novels, cult magazines and the fantasy worlds of television and the movies—until my best friend started telling weird tales of experiences that were very disturbing to her. My sense of what was real and possible changed almost overnight.

Until January of 1992, I never really believed in UFOs. Intellectually, logically, I knew that UFOs were probably real, too many people had reported seeing strange lights in the sky at night—all of them couldn't be misinterpretations of normal phenomena—and they couldn't all be crazy. The odds against our world being the only planet in the galaxy inhabited by sentient beings (with the most advanced civilization) are overwhelming. I've always considered myself to be open-minded, and am willing to suspend disbelief until I have enough information to make an informed choice. Besides, I grew up in the TV generation where news and fiction used to be separate events. I eagerly watched Neil Armstrong's first steps on the moon—I had no illusions that what I saw was real. It opened many avenues for my imagination to follow, when I had time—later. I watched *Star Trek* emerge as a science fiction cult phenomenon, but did not become a "Trekkie"—I had no illusions that what I saw was not real. I was fascinated by the creatures and the story lines; besides, Spock always appealed to me with his cold, calm logic in any situation. But to actually believe that we could work with (or against) sentient beings from other planets and galaxies within my lifetime—no way. A hundred years from now, yes. But not now. Please, not now.

I had intellectually accepted the possibility that UFOs were probably visiting the earth. It was an interesting phenomenon, one I hoped to see some day, but I never took the next logical step—that UFOs could be occupied. What little I knew about humans being in contact with alien entities was based on bits and pieces of movies, books and tabloid headlines. I hate scary movies; I never watch them all the way through. Most alien contact movies fall into that category. No, thanks, not for me. I prefer adventure, nature, comedy and the occasional psychological thriller.

I loved *Alfred Hitchcock Presents* and *The Twilight Zone* when I was growing up. The psychological twists in the plots kept my interest. No, I never saw *Close Encounters of the Third Kind* or even *E.T.* until a few months ago. I watched *E.T.* on TV and thought it was cute, but not realistic. I rented *Close Encounters* because I couldn't remember what happened and people kept asking if I'd seen it. I watched it and realized that the reason I couldn't remember what happened was because I had never seen the whole movie, only previews for it on television. I enjoyed the movie, although there were several places in it that made me very uncomfortable. (Too close to my own experiences?) The bright lights in the beginning at the railroad tracks made my skin crawl, and the views of the mother ship were disturbing, although beautiful.

My tastes in books run to murder mystery, science fantasy, James Michener, Dick Francis, Anne McCaffrey or technical books related to horse farms. Most of the magazines I'd read were horse related. Not a book or magazine in my extensive library had anything to do with UFOs or abductions. These were two subjects that I had never taken time to study. I considered UFOs as probably real, but anyone who claimed contact with alien beings had to be lying, crazy or just plain weird. Tabloid headlines proved it—flying saucer cults and little green Martians were all the product of demented or avaricious minds. I tended to stay away from people who were too far out of the mainstream of American life.

I still have a hard time believing that I am not in control of my life, that "little gray shits" have had a role in directing my life, and still do to this day. I'm frustrated and angry. How dare they! They never even asked my permission. I know now that I would never give it. They take my body and mind, perform unknown experiments on both, return me to this reality with hardly a memory, but with a body that suffers bruises, cuts, exhaustion, digestive disturbances and the occasional false(?) pregnancy.

I accept and reject their existence daily. I can believe in them when I know I have been abducted the night before, but that only lasts for a few weeks. When they become inactive for a month or so, I'm sure I made all this stuff up. I go back to denying that they are really abducting me. It doesn't change my belief that they are abducting others, I'm just not involved. Beth calls it my denial phase. I go through it continuously it seems. Whenever something happens to upset my comfortable, scientifically detached investigation of this phenomenon, I become panicked, or depressed, where I withdraw from everything and everyone around me until I can regain some equilibrium. I'm not sure I ever will really accept my involvement with the aliens; it's just too bizarre, too far from the reality I have known for all of my life. But, if I don't accept their reality in my life, I probably will go crazy.

Whatever happens in the future, I am going to continue to fight against the gray shits. I'll fight for my sanity, for my right to choose to live my life without interference from them, for your right to know what is happening to me and thousands of your friends and neighbors, and for abductees' rights to be taken seriously in their quest for physical, emotional and mental support in dealing with their personal alien invasion.

Now, I am forced to acknowledge the existence of UFOs and their occupants.

chapter 1

The Way We Were

1. Beth

The world as I understood it was to take a giant leap into the unreal, yet I was unaware of this impending transformation when I made a major life-changing decision in the spring of 1986. My parents, although elderly, were in good health and living comfortably in Colorado. My son, Paul, who was in the military and had recently married, would be moving with his new wife to Texas where Paul would be stationed for six months. I was then living on the east coast and realized that for the first time in twenty years I could go anywhere I liked. I had no responsibility to anyone but to myself and no ties to what had been our home for almost five years.

I had raised my son alone, having been divorced from his father since Paul was five, and had made every effort to provide a good home. I maintained a career that afforded a respectable income; we lacked for none of the essentials. We spent time together and enjoyed each other's company. I would miss him.

My feet were firmly planted on the ground, so when I found myself faced with this sudden freedom my instinctive reaction was to stay put, not make any impulsive decisions. I had worked hard to keep my home. I had planted beds of roses and spent many delightful hours tending them. Paul and I had built front and rear decks on the house a few years before and I looked forward to spending cool summer evenings relaxing there, enjoying my own company. That was the way it should be. I should be planning my time accordingly, savoring my middle years now that my son had moved away to start his own life. Yet with each passing day, I became more anxious, more impatient, as if I were expecting company that was determined to show up fashionably late.

Finally I decided something had to be done; I was stagnating and needed some diversion. I spent more time away from home, visiting friends and relatives, driving miles into the mountains just to enjoy the view, saving every good-weather weekend for horseback riding, which was my first love. Even so, I felt lost and undirected; I needed a goal to set my sights on.

On any typical workday I arose at 4:30 A.M. to start my day and was on the road by 5:30, ready and eager to begin. But on this particular April morning, as I plowed through the usual traffic congestion, watching cars jockeying for position in the increasingly heavy flow until a bottleneck brought us all to a complete stop, I found myself oblivious to these mundane events. It was as if my mind had detached itself from my body, spiraling out into the void. Suddenly I felt totally out of place in that mass of humanity and its machines, as if I were no more than a machine myself, following some ancient and obscure ritual which had lost all meaning.

Why was I doing this?

Because I liked my job.

Was it worth it?

Yes. The pay was decent and I had worked hard to get ahead.

Why didn't I just quit?

What?

Why didn't I just resign? Today.

What? What was I thinking? Why would I even consider such an idea? Surely this wasn't me talking!

And while you're at it, my little internal voice asked, why not sell the house and move?

This was just too much! Such an irresponsible attitude was utterly out of character and somewhere deep inside I recognized that, but I also recognized the truth of it. I was about to rock the boat significantly and, with any luck, I might not fall out and drown.

I never made it to the office that day. I pulled off at the next exit and returned home, feeling inexplicably free and light. By all rights I should have been horrified by the prospect of being unemployed, let alone homeless, yet I was seriously contemplating where I wanted to live after I sold the house! I was actually excited over the prospect even though I hadn't an inkling of how I might survive or of what my future would be like.

After arriving home, I immediately called the office and resigned, hinting that some personal tragedy had prompted this unexpected action, not the least guilty over the outright lie. By noon of the same day I had listed the house with a local realtor, called the Salvation Army and made arrangements for them to pick up furniture and appliances I had no intention of taking with me (wherever that might be), and notified my son and daughter-in-law that I was relocating and would inform them of my new address, as soon as I knew it myself.

By the first of May I no longer owned a house. But I did have a new job. My car packed with only personal belongings, I drove west to an uncertain future, away from the life I had thought I wanted, away from the security for which I had worked for so long, away from a "normal" life with normal expectations and dreams.

The employment I had found conveniently offered housing; if it were not for that one necessity, I may not have applied at all. I was hired immediately, after my first interview, though it was—and still is—a very odd turn of events. The position was that of recreational director of a summer camp for the physically and emotionally challenged. Not only was I unqualified for this position, I was also much older than desired and completely without related experience. Nonetheless, I found the job rewarding, the accommodations satisfactory (if somewhat primitive), and the campers and staff stimulating.

The position lasted through the summer months and in September I once again had to consider my next move. Less than a week before closing, another opportunity came to my attention: An employee of the camp had recently been hired as the manager of a racing stable nearby and had been authorized to hire another individual who could live on the premises and assist with the horses and facility. Since this was something I knew I was qualified to do, I accepted without reservation.

I had notified Paul of my plans and forwarded a change of address to him in Texas. Only then was I told that his wife, Sandra, was pregnant and would be due in December. I was to be a grandmother! This news thrilled me. My son, however, berated me royally for not behaving like an expectant grandmother; I shouldn't be acting like a vagabond, wandering aimlessly from job to job, living out of my car. I was hurt but had to admit he was right to feel his mother was behaving rather strangely. I couldn't explain my actions, couldn't begin to justify them. I only knew it was something I had to do.

The new job, though not difficult, required stamina and long days under less than ideal conditions. This was not a problem for me. I relished the physical activity and felt vibrantly alive. The problem came with the manager, who felt his position entitled him to certain liberties. I wouldn't tolerate the continual sexual advances, and so tendered my resignation. The abusive manager, who was married with two young children and obviously feared I might betray him to his wife, made sure I was not long in leaving. I returned to the bunkhouse later that same day to find my belongings stacked outside the door in the pouring rain. As before, I should have found my circumstances quite unsettling, but I did not. I calmly loaded everything into the car and drove into town, checking into an inexpensive motel until I could locate another job.

I had by this time decided working with horses was what I wanted to do, despite the hardships involved. My family was still aghast over my first move so I was hesitant to inform them of this latest development. Even then I understood I was a willing passenger on a train moving under its own momentum and I would ride it to the end of the line, wherever that might be. Something would turn up.

And, sure enough, something did.

While scanning the Sunday want ads the following morning, I noticed a small ad for barn help at a neighboring horse farm. The ad made no mention of housing, so I was reluctant to check it out. I scooped up the newspaper, went to breakfast and spent the next hour reading through the rest, my thoughts returning often to that first ad. Finally, having found nothing else that appealed to me, I returned to my room and placed a call to the horse farm. I was told that the position was still open, that housing was indeed available, and was asked if I could drop by that day.

Upon arrival, I met the farm's owner, Anna Jamerson, with whom I strangely felt a special bond. We were both certain we had not met before, yet we were both aware of some unspecified connection between us. This first impression proved more accurate than either of us suspected at the time. I was hired that same afternoon and moved myself and my few possessions into new quarters the following day.

It would be several years before Anna and I discovered our mutual pasts, years in which our lives, our perceptions and our beliefs would undergo drastic and irreversible changes. In a world too often occupied with its own diversity to permit lightning to strike the same spot twice, we had somehow found each other again. As if it were meant to be.

2. Anna

As I start to write this I would prefer to tell you that I had a normal upbringing, but I find I can't. Not that it was abnormal, but just not the stereotypical American family life; a nuclear family living in a small town, having the same group of friends to grow up with, graduate from high school, marry, have children and live happily ever after. I now know that few people's lives follow this pattern, and I am no exception.

I grew up as a "military brat" and travelled with my family whenever my father was transferred, spending no more than a few years in any one home. I looked forward to each move as I always knew that there were new and exciting things to do at the next duty station. I never became closely attached to people; I had few close friends, because I knew we would be moving soon. I looked forward to meeting new people. Another advantage was the opportunity to see the U.S., live in England, and travel to Europe during the summers. My mother and I were disappointed that my father didn't put in for another overseas duty station as soon as we returned from England. He settled his family in Florida so that we could have a more stable home life.

The closest I came to the American norm was living in one house from the time we returned from England until after high school graduation. I went to junior high and high school in the same Florida town, made more friends, had a few boyfriends, and settled down into the life of an athletically-active honor student. As graduation approached, I had dreams of marrying my high school sweetheart, settling down to a job that paid $150 a week (I'd be rich!) for a few years, and then concentrating my efforts on having and raising about a dozen children for the man I loved.

My parents, wisely, intervened and insisted that I go to college first, so that if anything happened to my husband, I'd be able to support myself and the children. Since I am a very logical person, that seemed like a good idea—besides I had received a scholarship to attend college and I'm not one to waste money. The rest of the story's short. I went to college, found a whole new intellectual world, was exposed to the Flower Children generation of the 1970's, joined them in parts of their rebelliousness and their seeking after explanations for every act by authority, forgot the high school sweetheart, and got on with my life. I never married, hold two college degrees (Experimental Psychology/Marine Biology and a Masters in Economics), have a good job doing analytical work for a natural resource agency, own my own farm, and until relatively recently was enjoying life and looking towards a future with increased prosperity and companionship.

I was going through a very rough time, financially and emotionally, in 1987. At work, I was shepherding a project that had tight deadlines, was behind schedule, too few workers (not too few staff, you understand), and I felt responsible for doing a perfect job for my contribution to the project—long days, intense pressure and lots of frustration. My home life was not much better.

Seven years earlier my parents, my sister and I decided to start an endeavor that my mother and I had dreamed of for years—our own horse farm. I still worked full time to pay my share of the mortgage and rode and trained horses in the evenings and on weekends. My parents fed the horses and cared for the horses during the day. My father did maintenance and the books, my mother fed the horses and treated minor injuries; I paid the bills. My mother even began to take riding lessons with me. Our dreams were beginning to become a reality. We struggled for years to make the operation work, let alone become profitable, but things kept going from bad to worse, with occasional bright spots—just enough to keep us going. In 1984, four of the horses died from various illnesses and accidents, all of which I blamed myself for; not doing a better job, not knowing how to prevent the deaths, attempting to do too much. But then came the five foals from the previous year's breeding decisions. They were better than I'd hoped for. So I got suckered in once again. We bought a bigger farm and kept the expanded horse business intact.

By 1986 the farm work had become too much for my parents to handle. My mother was diagnosed with pancreatic cancer, so I had hired a farm manager. He was a retired military colonel, knew about horses, and my father got along famously with him. Unfortunately, I was supposed to supervise his work, and we didn't see eye to eye on many of the details of raising and training horses. Evenings and weekends he did it my way, the rest of the week he did it his way. I was pretty adamant about not doing things just because "it's always been done that way." I wanted good reasons for doing things the right way. I had ten years of experience with horses at that point, read extensively, and practiced techniques on my horses and those of friends. I knew how I wanted my horses taken care of, but he thought he knew best. After too many arguments, he left in the spring of 1987, and I hired a young college graduate. When her friends went back to school in the fall, she gave me two weeks notice so that she could join them. My mother died the same week. I was not coping well. I could barely get through the day without crying or blowing up at someone.

I advertised in the local paper for a stable manager. I interviewed three people the first week, none seemed suited to me and my horses. Beth called on Saturday and my father set up her interview with me on Saturday afternoon. As soon as she drove up, I felt that I had met her before. That's not too uncommon—the horse world is sometimes very small. We talked as I showed her around the farm, and she passed my ultimate test—walking out into the pasture with 25 loose horses, remaining completely calm and enjoying it. She knew enough about horses and their care. Her most recent experiences were with race horses—no connections there. I had concentrated my time and efforts in dressage, eventing and hunter riding and a couple of local riding clubs. Beth was eager to learn, had no qualms about letting me know what she didn't know, and I was very comfortable with her. We thought a lot alike, we both had a love of horses, and there was that nagging suspicion that we knew each other, yet we couldn't pin down when we may have met. I offered her the job before she left. I was also desperate. She accepted. I was relieved!

Over the next five years we developed more than an employee-employer relationship. She became the best friend I have had since my college days. We had been gradually trusting each other more and more, exchanging more information on a personal level rather than just about the horses and the farm. The turning point was when I returned home from work one day and my best mare had just barely survived a very difficult foaling. Beth and our veterinarian had worked for several hours to save the mare; the foal was dead, but the mare was going to make it and eventually have other foals. Beth and I sat in her trailer all that night, talking, crying, smoking cigarettes and drinking wine to take away the pain. A few months later the trailer's heating and air conditioning system broke down for good. It seemed logical that Beth should move into the spare bedroom in the house. She did, and from then on she was a member of my family.

My feelings of helplessness were beginning to recede. I was regaining the control over my life that I so desperately needed. Everything was finally going right. It was time to celebrate new beginnings. Little did I know just how new.

chapter 2

Breakdown

3. Beth

I must be having a nervous breakdown, I thought anxiously. There could be no other explanation for what had happened that night, no reasonable explanation anyway. Nothing in conscious memory remotely compared to this imagined experience. For surely that was what is was; a flight of fancy, some outlandish trick of the mind, a flaw in the subconscious. I was, God forbid, having a nervous breakdown!

I sat huddled in the old rocker in the den, shaking and cold (even though that December night in 1991 was unseasonably warm), and fought against the obvious, wondering how I would explain what had happened that Sunday night without falling apart.

After having spent the weekend with my parents, who had moved back to the east coast a few years before, I was driving back to the farm around 8:00 P.M. It was a warm starlit night that seemed more like September than December, and the car windows were rolled partly down. I was only five or six miles from home when I was distracted by what appeared to be low flying aircraft approaching from the north. I didn't hear engine noise, which I thought odd since they appeared so low in the sky and too close not to be heard. The night, in fact, was unusually quiet—no other vehicles on the road, no lights on in any of the houses within sight, no activity, no barking dogs or yowling cats. This is a back road in a rural area and seldom has much traffic, but for some reason I was uncomfortable being there and wanted nothing more than to get home as quickly as possible. Even so, I was compelled to slow down and look more closely at these lights in the sky as they came steadily nearer, descending rapidly as they advanced. Checking that no traffic was in sight, I pulled off to the shoulder and came to a stop, leaving the engine running.

I was breathing rapidly and felt unduly concerned. What was there to fear? Just a couple of planes flying a little lower than usual. Why wasn't there any noise? Perhaps I was upwind and the sound of their engines was blown away from me. But there was no wind, and they no longer looked like ordinary planes flying low in formation. The lights now appeared to be attached to one huge aircraft. A jet, I wondered? No jet I'd ever seen could fly so silently.

Before I could decide what it was I was seeing, the lights suddenly brightened. I struggled to focus on the fuselage, trying without success to determine what kind of plane this was, but the lights were too bright to see past them. Resolved to settling this mystery, I turned off the ignition, got out of the car and walked out onto the roadway for an unobstructed view. I had to shade my eyes against the glare of these dazzling white lights, still believing I'd be able to identify them. They moved directly overhead and parked there like two giant headlights, so close I felt I could reach up and touch them. There was still no sound from these flying lights and the longer I stood there watching, the more frightened I became. Without warning, one of the lights zipped off to my right then hovered in its new location; the other remained in place. I strained to see what was behind this fast-moving light, but only succeeded in giving myself a headache.

Something was very strange here. I had a strong desire to turn around and run, as fast and as far as possible, but found I could not move from the middle of the road. Then remembering the other light, I glanced to my left, blinking to clear my vision.

Suddenly I found myself careening around the corner, five miles from where I was less than a second before, accelerating at a dangerous rate. The car threatened to flip over as it lurched onto its outside wheels, tires squealing ominously. I struggled with the steering wheel, trying to get the car back under control. I had no time to wonder how I had gotten there or why I was driving so recklessly. Miraculously, the car stabilized, all four tires hitting the asphalt with spine-jarring ferocity.

I brought the car to a stop, not bothering to move off to the shoulder. My stomach was heaving, my whole body trembling. I opened the car door just in time to throw up on the pavement. I had to calm down, pull myself together; I had to get home. I recognized the road. It led directly to the farm. I was only a mile from home and was sure I would be able to make it safely. I'd worry about what had happened after I got there.

Sitting in the den, a cold wash cloth draped over the back of my neck, I allowed myself to think back. I remembered pulling off the road when I'd seen the lights; I remembered standing in the roadway and trying to focus on the aircraft, but being unable to see past those lights; I remembered seeing the lights divide; I remembered turning to see if the remaining light was still there; I remembered blinking....

I could not remember what had happened immediately after that. I had no recollection of returning to the car, driving on down the road to the turnoff, or accelerating precariously around the bend. Whatever had happened from the time I was standing watching the lights until I found myself balancing on two wheels was a complete blank. I had either blacked out (but how could I have driven while unconscious?), or I simply experienced a loss of memory.

But why?

Like a lot of people, I had been known to daydream while driving on familiar roads or after many hours behind the wheel, sometimes not being able to recall anything I passed along the way. I have even drifted off into fantasy on occasion. But I could not ever remember being fully alert and out of my vehicle one minute, then becoming conscious of being several miles away driving erratically.

What had happened to me? If I hadn't blacked out, which seemed unlikely, I must have been moving under my own power; otherwise I couldn't possibly have gotten back in the car, started it up, and driven away. The only explanation was a breakdown of some kind. I had been—and probably was still—experiencing a nervous breakdown.

The reason for my behavior had a name, but naming it did not make it easy to accept. How would I tell Anna? According to her sister, Nancy, who lived with us, Anna had gone to bed about an hour before. It was quite late, after all, she had explained.

Late? I asked, surprised. It couldn't have been more than 8:30 or 9:00 P.M. I distinctly remembered looking at the dashboard clock when I pulled off the road the first time: It read 8:15. Assuming I had spent perhaps a total of fifteen minutes watching the lights then recovering from the speedy cornering, I should have arrived home no later than 8:30.

I asked Nancy what time it was. Quarter after ten, she replied, looking as worried as I felt.

This only served to deepen the mystery, and caused me to believe I was in a worse emotional state than I had at first suspected.

There was nothing to be done about it that night, at any rate, so I gathered what was left of my strength and dragged myself off to bed. By morning I would have decided what to do—and say—about the episode.

Perhaps it would be better to forget it ever happened.

Odds were it would never happen again.

I had not actually planned to tell Anna about the previous night's experience, but once I began talking I couldn't seem to shut up. I wasn't embarrassed, though perhaps I should have been. I was simply uncertain I'd be able to explain it in any coherent fashion. I was having enough trouble explaining it to myself.

I saw Anna as a pragmatic, self-assured woman who was accustomed to being in control—both of herself and her surroundings. This bizarre tale would be difficult for her to accept. To believe events had occurred as described, Anna would have had to adjust her thinking considerably. I was confident that she would approach this dilemma in her usual businesslike, down-to-earth manner; in other words, I could depend on her to believe as I believed: I had experienced nothing more extraordinary than a mental hiccup.

During the night (while I lay awake hoping to sort things out), I had concluded that the experience, although frightening, was probably not so unusual. I resisted worrying about my having had a complete breakdown. After all, wouldn't this irrational behavior have continued? By morning I felt almost normal; while relating the experience to Anna, I felt in control of my emotions and open to her insights, whatever they may be. I was not the least defensive. In the morning sunshine, the previous night seemed almost like a dream.

4. Anna

December 16, 1991 was the advent of my different reality.

Beth's description of her terrifying experience the night before left me feeling helpless again. Somehow her experience showed another crack in that structured world I'd tried to build for myself. I knew something was terribly wrong with her, yet I didn't have a clue how to help her.

This new twist, Beth's hallucinations, was something that had to be dealt with if my ordered life was to continue intact, proceeding towards definite goals. It was more than a "she's crazy, now who will I get to take care of the horses?" problem that I faced. Beth was a sister to me. I wanted to help her, I just didn't know how. Maybe she was going crazy; each person manages a finite amount of stress before they become dysfunctional, but a little time off should get her back on her feet and get her mind to stop hallucinating.

The Christmas season is usually stressful, so maybe after the holidays she'd be back to her old self again. I took two weeks off work so I could help around the farm and give her some time off to sort things out. Beth has always been passionate about her work and tends to do too much, get overtired, and then end up with pneumonia. After a few days in bed, some antibiotics and a good dose of boredom, she's ready to tackle everything again. Maybe the hallucinations were just her mind's way of telling her to take a break.

We'd all be back to normal soon. No drastic changes were necessary. By the new year, my life would be back in order.

5. Beth

I was startled when Anna, who had listened intently without interruption, suggested I might have seen a UFO and that the subsequent disorientation and memory loss may have been connected to that sighting. She reminded me of an unusual event in late summer of 1989 when Anna, her sister Nancy and I had witnessed strange lights behind the house. I recalled the incident clearly, but was surprised to discover that Anna and Nancy both described slightly different versions of what had happened that night.

It was around 8:30 P.M. and Anna and I had been watching a show on TV when Nancy called us out onto the back deck. We both went outside together, joining Nancy on the deck. She was pointing toward the west and seemed excited. I recalled seeing three very bright lights traveling toward us in a triangular formation. At first I thought the lights were running lights on a jet having recently departed from the airport (which was some thirty miles northwest of us), but if so, the airliner was flying too low on such a clear night and its altitude should have made it quite audible. Yet we heard nothing—not even the sounds of crickets or frogs, common night noises in the country. Could it have been a helicopter, or more than one helicopter? No, we decided. The sounds of rotors would be unmistakable.

After watching for several minutes (I recalled three or four minutes at the most), the lights halted about twenty-five yards from the house at an altitude of approximately two hundred feet. They remained stationary—and silent—for another minute while we watched curiously. Suddenly, one of the lights broke away from the other two and shot off due north, causing all three of us to cry out in surprise. Its departure was so swift, and unexpected, that I couldn't focus on it. All I could see was a trail of light—as if its shadow could not keep up—which lingered only long enough for me to register that I was indeed seeing it. The object stopped just as suddenly and hovered to the north of the house, where it remained for perhaps another minute before disappearing completely, this time leaving only a speck of a shadow as if it had sped directly away from us.

At this point, I recalled Nancy saying she was going to check around the front of the house to see if there were others, while Anna and I stayed to keep watch on the remaining lights, presuming these were also individual craft rather than one unit. When Nancy returned, saying there were no others out front, we witnessed the immediate disappearance of the other two lights. They did not fade away nor did they fly off; they simply vanished. I noticed the normal night sounds returning, a gradually increasing cacophony that suddenly seemed too loud.

Seeing no other activity, we all went back inside, Anna and Nancy going on to bed while I returned to watching TV. I don't recall spending much time analyzing what we'd seen, classifying it as unknown, though we had all considered the possibility that they were UFOs. We felt thrilled to have seen them and promised ourselves we'd check the papers for any reports of sightings in our area that night.

But now, Anna was telling me she recalled some of those events differently: In her memory, there were at least four lights, perhaps five. Three were hovering in a line about where I remembered them to be, and two more were to the south, one close to the main stable. She also recalled being outside for only a couple of minutes before going back to watch TV (though she readily admitted to witnessing the sudden departure of the one light which later flew off at tremendous velocity, as well as Nancy leaving us to check around front). Anna could not explain how she could possibly have been in two places at once, yet she insisted she had very clear memories of doing both.

Nancy had been listening to our conversation and added her own recollections of this bewildering phenomenon: She agreed with Anna that there were certainly more than the three lights I had reported seeing, though she hesitated to say just how many she had observed. She insisted she had never gone around the house to check out front, but had instead stayed out back with us the entire time. She did not believe Anna had gone back inside earlier, and denied having gone to bed when Anna did, but said she'd stayed up for another couple of hours talking to me about the incident. I had no memory of that late-night conversation with Nancy.

What a baffling mess! The three of us could not seem to agree on more than a few points. How could that happen?

Time tends to distort memories, I rationalized. Who knew what we'd seen that night. It was, after all, a long time ago. How we remembered the incident wasn't important, I asserted. It had nothing to do with what had happened to me!

Anna, however, seemed to feel it might. She proposed we contact someone who knew something about the phenomenon, perhaps an organization researching sightings of this sort. I was stunned. Only crazy people reported seeing UFOs, I cried, even if they are real. And for Anna to suggest we actually report the incident as being related to the subject seemed absolutely insane to me.

This was not the Anna I had come to know and admire! This was not the levelheaded pragmatist who believed only in those things she could see and touch. This was another side of Anna entirely; a brave, didn't-give-a-damn-what-others-thought person who was proposing that I go public with my psychological problems!

Initially I was confused by Anna's reaction, having hoped she would confirm my suspicions that I had experienced a mild—and temporary—mental breakdown, probably brought on by stress. Mental illness was treatable; certainly symptoms as obvious as memory lapse and hallucinations could be attributed to stress. It made more sense to contact a therapist than a bunch of nuts who investigated flying saucer sightings!

I'd had little interest in UFOs previously. There was one incident that occurred when I was about nine or ten years old that had never quite been satisfactorily explained, but most everyone could dredge up an event from childhood that could be classified as a mystery. I explained this to Anna, but she was resolute: What we needed to do was talk to the experts.

That was fine with me. Anna could talk to anyone she wanted. I, on the other hand, had no intention of talking to anyone. I already regretted telling Anna and wondered how I was going to stop this runaway train.

I had always been a bit unconventional, though I'd stopped short of utter eccentricity. At forty-three, I did not wish to be regarded by my peers as a weirdo who attributed psychological and emotional disorders to an invasion from outer space. Being considered crazy would be preferable to that!

chapter 3

The Fund

6. Beth

The Christmas holidays came and went with barely a word about my strange experience. January 2, 1992 ushered in the routine workaday world, reminding us that life goes on its merry way whether or not we choose to go with it. I had, for all practical purposes, put that December evening out of my mind, making an appointment for a business meeting later with a fellow board member for a therapeutic riding center which was struggling to get off the ground. Our business was to be combined with pleasure by meeting at a quiet restaurant about ten miles from the farm.

We enjoyed a pleasant meal and had fulfilled our business obligations in a little over an hour, departing the restaurant by 8:15 P.M. Since we had not seen each other during the holidays, my associate had presented me with a Christmas gift before parting, which she hinted was stuffed with goodies. The colorfully wrapped and ribboned box (about the size of a shoe box) was quite heavy, the aroma of freshly baked breads and cookies escaping, making my mouth water. I thanked her for her generosity and we got into our respective vehicles and left. I checked my watch again as I pulled out of the parking lot and saw that it was not quite 8:30. I expected to be home in twenty minutes.

At 8:45, as I approached an intersection only five miles from the farm, I noticed how deserted the area was. The parking lot of a convenience store located just past the traffic light at the intersection was empty except for one lone car parked near the front entrance. No cars were in front of the gas pumps, no people visible outside or in the store, from what I could see. There were no other cars on the road in either direction, no lights on in any of the houses along the route. This seemed rather odd, even for a weekday evening, as it was still early. Then again, I rationalized, it was just after the New Year, and presumably people were staying home recuperating from the holidays.

The light turned green and I cruised on through the intersection, unconsciously searching for signs of life. I felt as if I'd dropped into the Twilight Zone; it was so uncanny, so absolutely still out there. After having driven less than a quarter mile, I saw a bright light off to my left. It was traveling at low altitude toward me and appeared to be a private plane perhaps making an approach to the nearby municipal airport which was only a few miles behind me. I continued to watch it, slowing for a curve in the road, when I suddenly felt something was wrong. Some instinct in me set off an alarm; I knew what the light was. I became so frightened I thought my heart had stopped beating. Instantly the inside of the truck I was driving was engulfed in a bright bluish-white light. I closed my eyes against the painful glare.

When I opened them again, I felt a bone-jarring thud and saw that I was on another road, miles away from where I had been only a split second before, speeding around a moderate bend in the road. I struggled to control the truck's steering, which seemed reluctant to respond, easing off the gas until the vehicle was again under my command. My heart was racing, my stomach in knots, but I dared not stop.

I didn't know what had happened, couldn't recall how I'd gotten there, but at least I came to recognize the area and the road on which I now found myself: It was the same road—the exact spot, in fact—where I had witnessed the strange lights the month before, but more than eight miles west of where I was seconds ago! If I hadn't been steering around a similar slow curve to the right before my mind went blank, I would have surely run off the road when I came to and was initially unable to steer the truck.

How had I driven there without being aware of it? And why would I have gone so far out of my way? *And why couldn't I remember what happened?*

I sped home, arriving at 10:20 P.M., more than ninety minutes later than I should have. Leaving my purse, briefcase and forgotten Christmas present in the truck, I clambered out and rushed into the house, abandoning the truck in the middle of the driveway. Once safely inside, I collapsed into the rocker. Nancy came into the den, instantly sensing something was terribly wrong. Anna and Nancy's brother, Rick, who was staying with us for an extended visit, followed behind his sister to see what the commotion was about. Anna, apparently, had already gone to bed. I stuttered an apology for frightening them, but couldn't bring myself to tell what had happened. I didn't know what had happened—just as before.

Nancy hurried to pour me a brandy and I sipped it gratefully while trying to calm down and organize my thoughts. She and Rick stood by patiently until I was ready to talk about it. Gradually, I explained what I could recall of the drive home. I suddenly remembered I'd left everything in the truck and asked Rick if he would go out and retrieve my belongings. When he returned, he mentioned smelling a strong lemony scent inside the cab. I couldn't imagine what would have caused it, and didn't really care. I had enough to worry about as it was. At this point I became aware of a stabbing pain in my right earlobe and asked Nancy to take a look.

"You're bleeding!" she announced. "How did that happen?"

I shook my head, not having any idea, and reached up to touch the sore lobe. When I removed my earring, blood dripped onto my fingers and I held them up before me as if unsure whose blood it might be. Nancy brought out a damp washcloth, gently cleaned the earlobe, then leaned over to examine the wound.

"There's a hole here."

"I should hope so," I joked, surprised I had the capacity for humor. "My ears are pierced, remember?"

"I know that! But it's like someone shoved an ice pick through it. Let's have a look at the other one."

I had difficulty removing the other earring, soon realizing that it was a problem because the earring was inserted backwards, with the wire looped over the front of the lobe. I would never have put my earrings on backward; in fact, doing so would not only be painful, but nearly impossible. Once the left earring was out, Nancy checked the lobe and found it inflamed but otherwise unharmed.

By this time I was fairly composed, concentrating on thoroughly cleaning and disinfecting my ear lobes, and felt almost human again. This was so baffling! How would I ever figure out what had happened? I absently fingered the sore lobe, believing that having a cigarette would help me to think more clearly. I reached into my purse beside the rocker, but couldn't locate them, so asked Rick if he'd seen my cigarette case in the truck. (I knew there was a full pack in the case since I'd just opened a new pack before leaving the restaurant.) Rick had put my cigarette case on the desk and Nancy handed them to me.

When I opened the case to remove a cigarette, the case was empty. I stared at the cigarette case, confused. How could that be? The case had been closed tightly so they couldn't have fallen out, yet the empty cellophane pack was still inside the case! I certainly hadn't smoked an entire pack of cigarettes! I asked Rick if he had found any loose cigarettes inside the truck, maybe strewn across the floorboards or on the passenger seat. He said he hadn't noticed any, but would go back out and check. When he returned, shaking his head,

I borrowed one from Nancy, deciding to classify the missing cigarettes as just another part of the night's mysterious events.

I remembered the Christmas present, which Rick had placed on the desk along with my briefcase, and pointed it out to Nancy, explaining who it was from and that it held perishables and should be unwrapped soon. When she handed it to me to unwrap, I was startled to see its condition: The sides of the package were bashed in, the paper wrinkled and torn, and the ribbon gone. There were several strips of masking tape holding the edges together, tape I had not recalled being there when I received the present earlier that evening. But most noticeable was the package's weight. It was so light I was sure it was empty! Could I have unwrapped it and eaten everything in some uncharacteristic fit of hunger? That was absurd! I was actually ravenous, as if I hadn't eaten anything all day.

I opened the box and discovered a small plastic sandwich bag containing a half-dozen cookies and another foil packet holding four square-cut brownies. If these contents were doubled they wouldn't have been nearly as heavy as I knew that box had been when I last held it.

Rick explained that he'd found the present and the rest of my things on the floor of the truck, suggesting that perhaps the package had gotten damaged when it fell from the seat.

"Was the shock so great it lost weight in the fall?" I asked sarcastically, already impatient to call my friend and ask what was originally in the package. I didn't care if it was impolite to do so, didn't care how late it was. I had to know, for my own piece of mind, what was left of it.

Nancy asked if I wanted to wake up Anna and let her know what had happened. That was the last thing I wanted! Instead, I begged Nancy and Rick to keep it under their hats and not say anything to Anna until I had time to sort things out. If anyone told her, I would. When I was ready.

7. Anna

January 3, 1992 was the day I realized that things were not going to go back to normal as soon as I had thought. Beth had another incident the night before. Where was I? The same place I had been before—asleep! I somehow felt that I should have been able to help her, at least by being awake when she returned home.

This incident was somehow different than the one two weeks before. Beth had physical effects (cut ears, earrings in backwards), the truck smelled funny, and she was more emotionally upset than I had seen her before. She told me that she had screamed to herself, "Oh no, not again!" when she saw the bright light. She felt that she knew what was about to happen, but now had forgotten. Something more was going on here than just hallucinations.

I looked in the Christmas package and ate a brownie as we talked. Beth told me she remembered that the package had been neatly wrapped with a bow on it when it was given to her. The bow was missing when it was brought into the house. She remembered that the wrapping paper was sealed with masking tape when Rick opened it the night before, not sealed neatly as when it was given to her. As she talked, I grabbed a phone book and started looking in the business section under *Organization, Association of,* with no luck. When Beth asked what I was doing—looking for a UFO organization—she thought I was the crazy one! I don't know why I knew this incident was connected with UFOs, I just knew that it was. I finally located one under *UFO,* of course: *UFO, Fund For Research.* Why didn't I look there first!

I called immediately before Beth could stop me or I lost my nerve. I reached an answering machine! For the rest of the day we both tried to figure out what was going on, and were restless for our own reasons: I was anxious for the Fund for UFO Research (the Fund) to call back, but Beth was dreading the call.

In the early afternoon Don Berliner called and I nervously answered. I found I really had no idea what to tell him. After the standard disclaimer, "I know you may think I'm crazy, but I'm not," I told him that I thought my friend had been picked up by a UFO. Was that possible? Don was extremely supportive to my obviously confused and anxious babbling. He assured me that we probably were not crazy, calmed me down, and asked if he could speak to my friend. Beth was sitting in the chair, petrified, shaking her head no.

Don indicated that there were some reliable people around who investigated anomalous phenomena, and wondered if we would be willing to talk with one of them. After Don assured me that the person he was giving my name to was discreet, had dealt with several people that reported these types of experiences, I agreed to talk with Robert Swiatek. Beth said "maybe." By this time I had cajoled her into at least speaking with Don. After talking for a few minutes, she handed the phone back to me and I made polite goodbye noises. What had I just done?

Beth had gone to fetch the wrapping paper out of the trash, but never made it back to me. I heard a loud crash, quickly hung up the phone and ran into the dining room. Beth had collapsed in a ball on the floor, clutching the wrapping paper and sobbing hysterically. I hugged and consoled her until she calmed down. An image had flashed into her mind. She had seen the Christmas package being held by two very strange looking hands. Charcoal-gray hands, with long unjointed, but very flexible, fingers. It terrified her. Me, too.

Why did I obtrusively push Beth into talking with outsiders about these unusual events? Obviously, she was the one who thought she was going crazy, and it should have been her decision alone to talk to strangers about what she remembered. I did it because I knew she was in no condition to make rational judgments. I felt I was losing a good friend, and the only person I trusted to take care of my horses, to something that could be logically explained. I knew that I didn't have the knowledge or background to sort this out, but if we got in touch with some experts, they'd explain it all and we could go back to living a semi-normal life.

All this emotion just bubbling up all over the place was hard to deal with. No, it was *impossible* for me to deal with, that's why I was so adamant about bringing in professionals who could help us. I believed that this UFO kidnapping stuff was a rare occurrence; the experts could tell us what was happening, and then we could classify it as a great adventure to tell our friends and relatives. "I Was Kidnapped By A UFO, and Survived!" Unfortunately, it wasn't that simple.

Robert Swiatek called later on that evening and I briefly explained what I thought had happened to Beth. I again asked if it was possible for a person to be picked up by a UFO, yet not remember it. Rob wasn't willing to give me much information, although I think he did admit that UFO involvement was a possibility. I gathered some background information on who he was, what the Fund was and what it did, and how he might be able to help us. He offered me the names of a psychologist and another researcher he was associated with, Richard Hall, as references if we wanted to get more information before we talked to him in more detail. After getting his solemn promise to keep our names confidential, I convinced Beth to talk with him. Before disconnecting, Rob made an appointment to meet with us at the farm on January 18. I guessed that was when he was going to tell us the truth about abductions, and then we could get back to normal. Oh well, two weeks wasn't that long to wonder. Besides Beth seemed calmer, and I certainly was—I'd turned the problem over to an expert. Now I could relax. What a fascinating subject!

8. Beth

"I don't think I'm ready for this," I admitted to Anna after she'd made contact with the Fund for UFO Research (FUFOR). "Why would anyone seriously believe this has something to do with UFOs?" Anna had smiled sympathetically and, I thought, somewhat imperiously.

"Well, I think we need to talk to someone about what happened to you. You're obviously very upset by it, and unless you have a better idea...."

I didn't have a better idea. But the thought of talking to complete strangers—especially people who actually considered the UFO phenomenon deserving of a special fund for research—put me, I believed at the time, into a very awkward position. First of all, in relating what I recalled of the events that night, I would be risking ridicule, both from them (if it turned out my experiences did not relate to their field of research) and from Anna, who would be embarrassed for having contacted them when what I actually needed was a good therapist. Secondly, my being a confirmed skeptic prejudiced any expectations that this dialogue would be worth anyone's time.

"It can't hurt to just talk to him," Anna coaxed. She had placed the call earlier and was awaiting a call back hoping she could convince me in the meantime. I was still reluctant when the phone rang, so Anna took it upon herself to describe the event to Don Berliner, who apparently found sufficient relevance to suggest that he and an associate, Rob Swiatek, come by and speak with us personally.

This was more terrifying than the event itself! By the time Anna had relayed the conversation to me, I was in full denial and had decided not to participate in any way. If Anna wanted to rub shoulders with these fanatics that was fine. I would not embarrass myself by doing so.

By Friday afternoon, January 3, I had reluctantly agreed to speak with Don Berliner, figuring that once I had described the incident (what little there was to describe), he would put my fears to rest by admitting it could have nothing to do with the UFO phenomenon. I did not concern myself with how I would otherwise ascribe it, believing then that the breakdown theory was a more likely explanation.

Don listened courteously to my monologue, periodically asking for clarification on certain points, but receiving little of that since I had no conscious memory of the missing time. When I mentioned the wrapping paper and its condition, he asked that I save it so he and Rob could examine it when they came out to talk to us. Agreeing to do so, I turned the phone over to Anna and went into the dining room to retrieve the paper.

I found the wrapping paper on the bookshelf, certain that it was the paper in question, and pulled it out. I felt as if I'd been run over by a bus; I was weak, shaking uncontrollably, but unable to call out. I staggered back through the kitchen and into the den, where Anna was still on the phone with Don, and signaled to her that something was wrong. Still clutching the paper, I turned around and started back to the dining room, hoping I'd be able to make it to a chair before collapsing. The next thing I recalled (and only after I came to), I was crouched on the dining room floor, Anna supporting me from behind. I still held the crumpled wrapping paper in my fist.

I had relived a terrifying episode which may have occurred during the missing time of January 2. I was confused, frightened, and had trouble relating the memory (if that's what it was) to Anna, who remained seated behind me, encouraging me to calm down and tell her what had happened. I took a deep breath and began:

I had experienced a sudden flashback, and remembered being seated, leaning over with my arms resting on my knees. I felt pressure against the small of my back, as if someone's arm was supporting my back or keeping me bent over so that I couldn't sit erect. I was unable to move except for my eyes, but my vision was blurred and I couldn't determine many details. The air felt heavy and smelled organic, like a closed greenhouse on a warm summer night. I caught motion out of the side of my eye and saw two hands holding a colorfully wrapped package. I recognized the Christmas present as the same one my friend had given me that night. The package was still wrapped as I remembered, including the bright red ribbon and bow.

The hands holding the box were so startlingly unusual that I couldn't accept what I was seeing. They were charcoal-colored, very smooth, and appeared to have only four fingers—and no thumbs. The fingers were very long and thin, seemingly without joints. This image so frightened me that I had tried to scream aloud, but could only gasp for air. I thought I was going to die, almost wished I would. I seemed to have no control whatsoever! I tried again to scream, and saw a hand before my eyes; two fingers closed over my lids, forcing my eyes shut. For some reason I was no longer afraid, and felt as if I might even fall asleep. I became so completely relaxed it no longer mattered where I was or what was happening to me. I simply didn't care.

After describing these images to Anna, I began to calm down, feeling more in control of myself. I glanced at the balled up wrapping paper, flattened it out in my hands, and frowned. It wasn't even the right paper. I tossed it down and struggled to my feet. I went directly to the waste basket in the living room and reached inside, withdrawing the "right" wrapping paper and smoothing it out. I studied it for some time, but no associations emerged.

What was going on? What kind of creature had hands like those I saw? Was what I experienced an authentic flashback, recalling a real event, as it seemed to be (it was so vivid)? Or had my overstressed mind created the whole episode? Was I really going crazy?

Could be, but I wouldn't be able to rest until I had spoken with my business associate and asked her what had been in the Christmas package she'd given me that night. I placed the call, going over in my mind how I would approach the subject without sounding either ungrateful or weird. After the expected pleasantries, I blurted out the real reason for my call. To my immense relief, she laughed, apologizing for giving me a present of food which wasn't recognizable! I laughed with her, but insisted she tell me what had been in the box.

"Two small fruit cakes, a dozen brownies, and a dozen cookies," she announced hesitantly, finally daring to ask why it was so important to know. "Isn't that what was there?"

I paused, wondering how much I should say. "Well, no. Not exactly." I told her what we'd found in the box, not surprised that she should be confused, too. I did not mention the condition of the box, the missing ribbon, or the masking tape; nor did I feel comfortable revealing details of Saturday's flashback, telling her instead that Rick must have dug into the stash for breakfast!

I was beginning to feel frightened. I knew I hadn't opened the present or eaten any of the contents between leaving the restaurant and arriving home. I was sure I wouldn't have consciously driven so far out of my way. I knew I hadn't put my earrings in backwards. I knew I hadn't smoked a whole pack of cigarettes, although Anna jokingly suggested that stealing cigarettes might be the only way aliens could get them since it would have been awkward for them to stop at the store and buy some!

I wasn't so sure of my mental state. I wasn't so sure Saturday afternoon's flashback wasn't imagination. I wasn't sure of anything!

9. Anna

On January 18, 1992, I waited impatiently all morning for the investigators from the Fund to show up. I wanted this enigma solved, and they were going to provide the answers. About one o'clock in the afternoon Don Berliner and Rob Swiatek arrived.

After introductions, we settled in the office for a short visit. They were both very courteous, seemed like they knew what they were talking about, and most important—for me—they seemed to be approaching this from a scientific point of view. Notwithstanding their investigatory role, they seemed concerned for our emotional well-being, and they listened without judgment! Not once did I get the feeling they thought we were crazy. I found Don and Rob to be open-minded people, concerned, but distant. Sharing information about the Fund, their backgrounds in this type of research and our backgrounds helped to reassure me that I wasn't dealing with a bunch of fanatics.

Rob asked if he could tape record the session since that was easier for him than taking notes. We agreed, and soon were almost comfortable talking with Don and Rob. (Unfortunately, the tape recorder malfunctioned, and all Rob now has of our first meeting is two tapes with a lot of static on them.)

They had planned only an initial interview with us that day and then to follow up with a more in-depth interview about Beth's experiences. But, Beth had returned from Pennsylvania on January 12, where she had been teaching a riding clinic, with a large triangular burn on the back of her right hand. Strange things were happening too frequently, and I had to know if they were connected. My first theory was rapidly being demolished: Normal people can be kidnapped by aliens once, maybe twice, if you were real unlucky.

As we became more comfortable, Beth told of her memories of the strange events that seemed to be plaguing her. I asked occasional questions, but mainly sat in the background, absorbing information. Beth told them about the lights of 1989 in the backyard, and what she recalled of the three incidents in the last month. They looked at her burn, at her earrings, and at the wrapping paper from the Christmas package. I told them what I remembered of the 1989 lights, and couldn't understand why Beth and I remembered it so differently. It had been several years, but even then I couldn't see myself coming inside the house and watching television when there were UFO lights in the backyard. I found myself crying at times as Beth described her terrifying memories. At one point during the interview she had a memory flashback (sitting, leaning over her knees, with a gray hand in front of her face), and I held her as we both cried until we could calm down enough to talk to Rob and Don again. I'm sure they thought we were overwrought, emotional women!

I did find out a little bit about abductions from Rob and Don, but mostly it was frustrating. If I asked a question about the phenomenon, usually they would respond by saying that "it may be possible," or "I've heard of that being reported." They didn't volunteer anything but the barest information about abductions. I wanted answers. It wasn't until much later that I realized part of their reluctance to answer my questions was because there are no definitive answers.

I reluctantly left the conversation about 5:00 P.M. to feed the horses (I was already half an hour late). Rob and Don were gone when I came back in the house. Before they left, they encouraged us to contact them or Richard Hall at any time if we had questions, and offered to find a hypnotherapist should we want to recover other memories of the events. They asked us to keep a personal journal of our feelings, memories and events. A journal could help to put things in perspective and to process and work through the trauma of the experiences.

Did you notice that I am now using "we" instead of "Beth"? One of the most disturbing things for me to come from the initial conversation was Rob's and Don's suspicions that I might not have been just an innocent bystander to Beth's mishaps. They felt that I had reacted with too much emotion to some of Beth's narratives, and at least once (Beth's description of her flashback), I had reacted to Beth's pain before she did! They had also noticed our tendency to finish each other's sentences, say the same thing at the same time, and speak to each other with incomplete phrases that obviously did not obscure our understanding. They asked if I had any psychic abilities—of course not! That's for people on the fringe of reality. I was firmly rooted in the here and now. They were also concerned about the discrepancies in Beth's and my accounts of the 1989 lights. More may have happened than just seeing a few strange lights over the barn. They cautioned us not to read magazines or books about the phenomenon; other people's experiences or speculation of what was happening might confuse, disturb or terrify us, or contaminate our memories.

They left two very confused women that day. I somehow felt better and worse at the same time. I had the feeling that this UFO kidnapping business was likely to happen again, but the field of study looked to be more fascinating than I had ever thought possible. It looked like some of these people I had labeled as religious fanatics or screwballs may have been telling the truth to an unbelieving public.

10. Beth

My nerves were dangling from a thread. Don Berliner and Rob Swiatek, though compassionate, disappointed me. They were supposed to reassure us! Anna and I had just about concluded these incidents must have a logical—and *earthly*—explanation, yet the Fund's representatives said nothing to validate our conclusions! In fact, they were unexpectedly guarded in their opinions and responses to our questions about whether these experiences were unusual.

During our conversation, while Rob's tape recorder sat unobtrusively on the edge of the desk, I wondered what these men were thinking, wondered what they were looking for. Finally, unable to restrain myself, I asked Rob if anything like what I'd described had been reported by other people. His response, though vague, told me the one thing I didn't wish to hear: that the missing time episodes, flashbacks, panic attacks and unexplained wounds were common complaints from those whom they had interviewed. When asked specific questions, such as, "Has anyone you've spoken to described strange looking gray beings with four fingers?" their standard response was, "We've heard of similar descriptions." Their answers were always frustratingly ambiguous, leaving me feeling almost foolish, as if I were being interviewed in order to obtain more ambiguous information to add to the Fund's collection. Where would all this lead? Could these people help us sort this mess out or not?

I didn't wish to be rude, or sound ungrateful, so I answered all their questions to the best of my ability, all the while wondering what it would take to get answers to some of our questions. After going through two tapes, the recorder was finally turned off and I visibly relaxed, as if the tape recorder had been judge and jury over my narrative, and therefore, over me personally. I didn't so much care if I was believed; I reported only what I remembered experiencing, not whether I felt it was imaginary or real. I would have preferred being told my remembered experiences were unheard of, nothing even remotely similar having been reported before, and that I should see a professional and try to determine what underlying trauma had caused these hallucinations.

Unfortunately, this advice was not forthcoming, and as our discussions continued, I came to the realization that my experiences—no matter how strange—were not unique. This awareness was intensified by Rob's caution against our reading any books, magazines or other literature relating to UFOs or abductions to reduce the chance of our memories being distorted or our being influenced in any way.

I suddenly recalled having bought a paperback several months before (but couldn't then recall the title or author) and having been unreasonably disturbed by the book's cover art, so much so that Anna and I actually *burned* it! When Rob questioned me further about this book, asking that I describe the cover, I had difficulty doing so, still seeing it clearly in my mind's eye and still upset by that image.

"It was a face, but not a human face," I explained. "The skin was grayish and smooth, and the eyes were huge, black and slanted upward, sort of almond-shaped. The chin was long and pointed, and the forehead was too big. The whole image was frightening, and I just couldn't read the book because I kept being drawn back to that picture."

Anna interrupted as she recalled the book in question. "That was *Communion*," she said. "I remember neither of us wanted to read it, but I don't remember anything about burning it!"

Since Anna was not a horror book fan, and I was (I'm a true Stephen King fanatic), I hadn't been surprised by her rejection of the book. What surprised me was her initial reaction when I offered it to her. She had refused to even touch it. I described Anna's reaction to Rob and Don, expecting them to shrug it off as personal preference.

"That was Whitley Strieber's book," Rob interjected, "I remember the cover art and I can see why it was upsetting to you, considering these recent events." Yet he wondered about Anna's denial of having participated in the book burning.

"But you were with me," I insisted, turning to Anna and hoping she would remember. The memory of it was so clear to me. I couldn't have been mistaken! "We went out together, put it in the burn barrel, and set it on fire! Don't you remember?"

She didn't, and I again began to question my memories. If this was a false memory, how much of the rest was false, too?

Don tried to console me, explaining that some memories were more reliable than others, and that some people retained specific memories (if they felt they were important), while others relegated the same experience to storage because the event was not important enough to keep on the forefront. This made sense to me, but it didn't ease my mind. I still questioned my need to remember this one seemingly trivial incident so vividly, while Anna, who's reaction at the time was far more intense, had little memory of it.

This enigma was becoming more and more difficult to grasp, and although I appreciated Rob and Don's patient understanding, I felt their visit had accomplished next to nothing. I still didn't know what had happened to me, still didn't believe my experiences were

related to UFOs or abductions by aliens, and still felt the help I needed could not be found with the Fund for UFO Research.

chapter 4

The Pendulum

11. Beth

Seeing is not necessarily believing. Though we had been advised by Rob and Don not to indulge our curiosities by reading UFO-related material, Anna and I could not resist doing some amateur investigating of our own. Even so, we discovered our built-in belief systems were well established, and found it difficult to accept some of these accounts, no matter how well presented. Perhaps it was because we didn't want to believe. After all, by accepting the possibility of truth, we would need to greatly modify our perceptions of reality.

I, for one, could not imagine myself as a member of this exclusive club for alien abductees. I could accept—even believe—that UFOs exist since too many responsible (and *sane*) people have reported seeing them, but I was still unwilling to embrace the alien abduction theory. It wasn't that I suspected these "victims" of creating a hoax; I truly felt they were poor misguided people who had seen or experienced something unexplained and their minds had filled in the blanks with fantasy and false memories. This self-defense mechanism made more sense to me than that ordinary people were routinely being taken hostage by aliens.

In the "real world" if an offender was tried and proven guilty of kidnapping, he would be subjected to harsh penalty, likely imprisonment for up to thirty years. So, I reasoned, how could this kidnapping by aliens be going on right under the noses of law enforcement? If the claims of abductees were justified, why weren't the authorities notified and action taken to protect the victims of these assaults?

Later, as my understanding of the abduction phenomenon expanded, I realized all my previously accepted strategies for dealing with such violations no longer applied. It had been tried before by other victims, resulting in humiliation, embarrassment and frustration over law enforcement's inability to take these reports seriously. And I could hardly fault them for their reluctance to investigate these claims. From my own scanty memories of events, I could appreciate how difficult it would be to report an abduction—after the fact.

First of all, there was the problem of believability; the mere description of these abductors would be enough to classify the victim as deranged. Second, explaining how the kidnapping was accomplished (often while in a vehicle, in bed or behind locked doors) stretches the credibility of the witness to extremes. How did the offender force the victim to stop his or her car on a lonely road, get out of the vehicle and go with the abductor, and then be returned safely if the victim was not a willing party to the abduction?

If the abduction took place in the wee hours of the morning while the victim was in bed, how had the abductor(s) gotten inside if not invited? And how would one explain a bedroom abduction succeeding despite the presence of the victim's spouse, sleeping undisturbed in the same bed? How would a victim of one of these abductions explain being floated through locked doors, closed windows or solid walls? If I was having trouble believing what little I recalled of these strange events, what was the likelihood of the police accepting them?

It was time to reevaluate my position. I would need to set guidelines as to what was acceptable for me and what was not. These guidelines were not to be set in stone (even early on I realized that my horizons were continually expanding), but they were necessary if I was to function in this reality. The more I could assimilate, the easier it would be.

But I couldn't seem to assimilate. Nothing in my experience had prepared me for having to blend two such unrelated realms. If I could accept one as the true reality while packaging the other as a curious (yet unproven) phenomenon, I might find it easier to cope. It was extremely important to me that I continue on with my life with as few uncertainties as possible. As I saw it, living day to day was enough of a challenge. So I learned to compartmentalize instead.

This prepackaging worked well, for a while, until things began to pick up. Interference in my recently accepted reality became more frequent, again presenting me with a dilemma for which I was less than prepared.

12. Anna

I became concerned that I may also be experiencing abductions. It was fascinating to think that I might be involved in this heady experience. It might even be fun if it weren't frightening. I'd seen Beth's panicked reaction to her abbreviated memories and I didn't want to go through that. But even if I were, I was stronger than she was; after all, hadn't I survived by taking control of my life and ordering it pretty well?

Even though logically I knew that I was unique, I still couldn't help believing that what I saw and felt was very much like what everyone else went through. So, if things happened to me, I assumed similar things happened to others. I am still shocked to find out that is not true in many cases—but I still can't help believing it.

I remembered that Rob and Don had suggested we both start keeping journals of anything strange that happened, or any odd things we remembered. I started keeping one on January 21, 1992 and much of what I disclose to you comes from my journal.

I started questioning all those little things that seemed odd in my life, wondering if they were in some way connected with UFO activity. I thought not, but wrote them down in my journal anyway. Keeping a journal is very therapeutic—I get to talk to myself and examine fears and feelings that I had always put away for another time. All those odd little things that I explained away, yet really wasn't completely comfortable with. It was time to examine all those little events stored away in the compartments in my mind.

I still didn't know much about the abduction phenomenon, but knew that usually there were periods of time that one couldn't account for. I'd had several of those when I was driving a car, but everyone gets highway hypnosis. Throughout my life I've always liked driving across country, done a lot of it, and preferred back roads to freeways. But there were several incidents I remembered where I would be on the same road, farther on down the line, with no memory of the intervening twenty to thirty minutes. I'd always put it down to daydreaming, leaving just enough memory behind to operate the car safely, but I seldom remembered what I'd been thinking about. Could these have been abductions? Probably not. Anyway, not all of them.

There was one disturbing incident that happened in the summer of 1990. I had been driving from Alabama down to Florida. I'd decided to take back roads to avoid the freeways since I didn't have an appointment until the next morning. I'd been told it was about a three hour trip; I figured four hours on the back roads. I left Alabama about 4:30 in the afternoon and arrived in Tallahassee at 10:30 that evening with a bad case of the flu. I'd been given a speeding ticket in Georgia, so couldn't figure out why it took so long. I was so sick I really didn't care. All I knew was that I had to recuperate before my meeting the next morning.

When I was a child, I used to fall out of bed a lot. I never really got hurt, except once when I cut my head on my sister's shoe buckle. My parents would hear a thud, check on me, and put me back in bed—I slept through it. The strange thing I'd never thought about before was that I was falling out of the top of a set of bunk beds! My parents assumed I was a very hard sleeper. Was that connected? I don't know.

When I was a small child, probably the summer of 1952 or 1953, I was lost on a beach in New York for some time. My mother was frantic when she found me. I didn't know I'd been lost.

I never thought of myself as having sleep disorders, and I still don't, but there are several things that I found strange upon waking some mornings. These things happened occasionally, but not often enough to worry me. I just made a quick semi-plausible explanation for each one, and then put them all into that box in my head for weird things that someday would be explained.

Logical explanations for illogical events:

I'd wake up with my nightgown on inside out or backwards when I was sure I had put it on right the night before—I must have done it wrong the night before.

I'd wake up with my head at the foot of the bed, my feet on the pillows—gee, I must have been real restless.

I'd have black, oily, organic stuff under all my fingernails—I didn't think there was any place on my body that was that dirty, but I guessed it was past time for a shower.

I'd wake up with a nosebleed—must have been picking my nose at night. I was always getting nosebleeds that way in the daytime.

My hair would be in snarls when I woke up—must have tossed and turned a lot.

My feet would be very muddy, or I'd find dirt in the bed—I really should remember to take a shower more often!

I'd wake up with a tremendous earache—my mother told me I'd slept too hard on that side of my head. Sure enough, within a few hours the pain was gone.

Occasionally, when I had my period, I'd wake up in the morning with blood all over the sheets because I didn't have a tampon in—must have forgotten to put one in. What was more troublesome was the mornings I'd wake up and remove two tampons!

When I was in high school and undergraduate school there were several times (eight or ten?) when I had found fingerprints on my upper arms; one print on the underside and two or three on the top. I bruise easily and assumed someone had grabbed me. I never remembered the incident that caused them. I am physically active and always seem to have bruises on my body, most of which I remember getting. Every so often I would find a big bruise on my arm or thigh, one that should have hurt a lot when it was done, yet I don't remember how it happened. I'd always tell myself to remember more carefully next time. It didn't work. I still get large bruises I can't account for.

Growing up I suffered all the normal childhood accidents, diseases and mishaps. I do remember having lots of bloody noses, earaches and stomach aches, more so than my brother and sisters. I used to get nosebleeds all the time, even wake up with them. My nose always seemed to be sore, but I always ran around at top speed and fell out of trees a lot. It bled even when I hadn't had a finger near it! The earaches were never severe, but the stomach aches sometimes were. I remember several times, while we lived in England, being rushed to the base hospital in the middle of the night with stomach cramps. The doctor always gave me some foul tasting stuff, said it was just an upset stomach, and sure enough it was fine by the next morning.

Many abductees have scars that they can't explain. Before 1990, any scars on my body could be accounted for, except for a couple of small indentions near my knees. But as a kid I was always falling down, skinning my knees, or stubbing my toes into bicycles, curbs or even just the ground. Sometimes I think I was a real klutz in my younger days.

Watching a television commercial, I had another memory of an unexplained thing. I was getting my Masters degree when it happened. I'd had a fender-bender and had a loaner car from a rental agency. I was riding horses at the time and had visited the stables that day. It must have been in the spring of 1979 because it was so muddy in the stable yard that there was a Volkswagen that had to be pulled out that same day. The next day the rental car wouldn't start. I called the company—they picked it up and gave me another car. The man from the rental agency called that evening and asked me what I had been doing with the car. He was convinced that the car had been driven through plowed fields! The undercarriage and the engine were so coated with dirt that it wouldn't start. I explained that there was mud at the stable, but that it wasn't that bad. I don't remember the outside of the car being more than normally muddy. He was adamant that someone had driven it through a cornfield, or some such thing. I hadn't driven the car the previous evening, having stayed home alone, and no one had borrowed the car. I didn't remember anything strange having happened; I barely remembered that evening at all. Nothing stood out about it. I assumed the mechanic was just the excitable type and the stable mud was more than he was used to.

On February 3, 1992, Beth had a compulsion to buy some UFO books. That evening I looked at the cover of the Vallee book she had bought and allowed my mind to open. I wanted to find out if I felt anything. I felt a little apprehensive and then shuddered before I gave it back to her. The power of suggestion at work? I thought I should feel something? I didn't know what was real anymore and what I thought I should be feeling—empathizing with Beth? I didn't know. I'm still not really sure. I had been toying with the idea of reading these type of books all that month—just never did anything about it.

I finally read Budd Hopkins' *Missing Time.* I found it hard to get started, although the words on the page were just normal words, I really had to concentrate to keep going. I also had the feeling that I shouldn't be reading it. Was this from Rob's and Don's caution not to read about UFOs if I planned to try hypnosis? I hadn't made up my mind about whether to try hypnosis. The things described in the book were so very similar to what Beth had remembered that it was frightening. Maybe this stuff really was real. Once I got through the first part I couldn't put it down. It was frightening for me in that it clarified the idea that I could have had an abduction experience, yet not remember any of it.

I tried to read *Transformation* by Whitley Strieber. Read the first few pages and had to put it down. Yuck! I have never liked to read about blood and guts and gory things and the first few pages seemed to indicate that this was what the book was going to deal with. I've never liked to deal with scary things in my reading—give me a pleasant science fantasy or murder mystery, not Steven King! I guess I was not ready to accept that the aliens were malicious and terrorized people on purpose.

I began having panic attacks—getting scared sometimes when I was out alone in the evening riding my stallion or driving home alone after dark. I started noticing that street lights seemed to go out as I passed them much more frequently than other people reported. As I'd drift off to sleep at night, I'd be momentarily terrified by images of a flying saucer or huge black eyes two inches from my face. Then there was the figure I saw in my room in the middle of the night (3:45 A.M.). I wasn't panicked. I just turned on the light and there was no one there. Only later did I learn that the electricity had been out in Beth's and my rooms (only) for half an hour at the same time as I saw the figure. She'd had strange dreams that night. And then we found that we were both dreaming of the grey beings on the same nights.

Then there was the time I knocked on Beth's door at 4:00 A.M. and told her I was going back to bed. I don't know where I had been, but my feet were muddy. I had no memory of being outside. I still was not convinced I was involved—I had just been reading too much. I still had lots of logical arguments for anything that might be happening.

I began calling Rob a couple of times a week and reporting what was happening, asking for explanations, and needing reassurance that everything was okay, even if it wasn't normal for most people. Talking with Rob was frustrating—he never volunteered information. I think he was afraid of leading me (us) to report things that only my mind thought had happened. If I knew the right question to ask, he always gave me a straight answer, or at least said, "That's been reported before." I started to rely more on Rob with each passing month, but he seemed to be overwhelmed by everything that was happening. I kept getting the feeling that this was the most complex case he'd dealt with, and some of the questions I was asking he just couldn't answer. I felt I was frightening him more than I was frightening myself!

Beth couldn't offer me all the emotional support I needed either. She had enough of her own stuff to sort through. Even though we talked constantly, my more reserved personality didn't allow me to talk as much as I needed to. Beth still could not bring herself to talk to anyone but me and I couldn't handle it by myself. Without Rob's support, I could not have handled all this emotional trauma as well as the new world view that I was beginning to conceptualize.

Such a problem. Beth wanted me to be involved, or knew I was subconsciously, and needed someone to be like her. I had this terrible need to deny that anything was happening to me, yet in the back of my mind I thought it would be fascinating to be involved and have my own experiences to relate.

13. Beth

Things are getting more and more curious. On March 8, 1992, I seemed to have lost another hour, and if not for Anna I most likely would never have known about the missing time.

I had been returning home after visiting with my parents over the weekend when Anna's car, which I had borrowed, suddenly stalled on an incline, all the dashboard lights coming on at once. I quickly glanced at the clock (though I don't know why I felt it was important to do so) and saw it was 7:00 P.M. The car was steadily losing momentum, and even though it was a standard transmission and I tried to restart the engine by popping the clutch, the engine refused to kick over. It became urgent that I get the car off the road and onto the shoulder before all forward motion ceased. I managed to do this just as the dashboard lights winked out and I breathed a sigh of relief to be clear of the road. There was very little traffic, as the route was mostly used by local residents or semi trucks cutting across between two major highways. I wasn't yet concerned, believing someone would surely stop and offer to help before too long.

Since the headlights were still on, I decided to pop the hood and see if something vital had worked loose causing the stall. But with no oncoming traffic so far (and no flashlight) I was unable to see into the engine compartment and finally gave up trying. The headlights would be no more useful than a warning to oncoming traffic that I was there. I got back in the car and decided to save the battery by turning off the headlights and putting on the flashers instead. After sitting for some time (about fifteen minutes), I noticed something move in the roadway ahead. It appeared to be a turkey! What on earth was a turkey doing in the middle of the road? I wondered. If a semi came hurtling down the hill, it would pulverize the poor thing. As if following orders, a tractor trailer did indeed appear at the crest of the hill, bearing down on the turkey. Suddenly the bird just disappeared! The truck could not possibly have struck it since it was still several yards away, but the turkey was no longer there and I never saw it go.

(Note: This seemingly trivial sighting of a turkey was later determined to be a possible screen memory, or perhaps a signal of an event unfolding. Similar sightings of "out of place" animals have been reported by other experiencers who have expressed a belief that the animals herald the appearance of aliens or a disruption in the abductee's physical perceptions. I have not been able to determine what the turkey represented, if anything. It may have been nothing more than a real turkey that found itself in the wrong place at the wrong time.)

Soon after the semi passed, I saw a white sports car drive slowly by, turn right at the base of the hill behind me, then pass me again going the other way. It returned several times within a few minutes, each time slowing as it passed, finally pulling in behind me with the car's high beams blinding me through the rearview mirror. The car's engine was turned off and I heard a door open and close. Not being able to see the driver through the rearview mirror, I glanced into the sideview and saw a tall man wearing a huge Stetson hat walking toward me. He appeared to be carrying something in his left hand—perhaps a flashlight. Realizing that this could be a dangerous situation, I quickly locked all the doors and rolled up the driver's side window, leaving an opening large enough to talk through.

The man stopped at the driver's side door and asked if I was having trouble. I briefly explained what had happened and he offered to take a look under the hood. I saw him shine a light under the hood, noticing that he wore all white and that the clothing was very snug-fitting. Then I heard a voice that sounded uncomfortably close order me to start the car. It was not a request and I did not even hesitate to do as told, even though I knew the car would not start. It didn't.

The man then moved around to the passenger's side of the car and leaned down. I saw he had a head that seemed too large for the rest of his body, a mustache that covered most of his lower face and that ridiculously oversized hat, that as large as it was looked too small for his head. I began to feel uncomfortable and moved back against the driver's side door, wondering if I had reason to be afraid of this person. The man then retreated, moving behind the car and around to the driver's side once again. He suggested I walk across the road to a house he pointed out to me, saying he was sure they would let me use the phone to call for help.

I waited until the man had left, then got out of the car, locking it behind me, dashed across the road, and picked my way up the gravel drive toward the log house on the hill. Only a porch light shone from the house and it seemed to take forever to reach it as I was uncertain about the footing in the dark. When I reached the porch and knocked on the back door, I was met by an exuberant black Lab and his mistress. Before I could explain why I was there, the woman opened the door wide and ushered me in. I asked to use the phone, saying my car had died right across the road from her house. She led me upstairs, showed me the phone and I hurriedly called Anna to pick me up.

Heading back to the car, I noticed the driveway seemed much shorter than it had going up to the house, but assumed that was because I was more familiar with it and my eyes had adjusted to the darkness. (The porch light had been turned off the moment I stepped outside.) Checking both ways before crossing back to the car, I saw lights at the bottom of the hill and realized I had wasted a lot of time; the small village of Delaine, which contained a gas station, post office, a few private homes and a country store with a public phone booth was nestled just around the corner! Why hadn't I just walked on down the hill and called Anna from the public phone?

Anna came along soon after, passing me to turn around in Delaine so she could pull in behind me without blocking traffic (which had increased considerably while I waited for her arrival). While riding back with her, I related the evening's frustrating events, laughing about the man in the Stetson hat driving the white car. Anna stared at me quizzically. "That's strange," she remarked. "When I turned around in Delaine I saw a man standing by a white sports car, but I didn't pay much attention. When I got to the stop sign I looked in my rearview mirror and saw the car pull in behind me. I remember thinking that I have to watch which way he turns when we pull out. But it's really strange," she thought aloud, "I forgot to look! When it was clear, I just pulled on out and never gave it another thought until you reminded me!"

It was strange, but more odd still was the realization that I had lost over an hour. I was certain I could account for the time I had spent in the car waiting for help (even though the dashboard clock had died with the rest of the indicator lights), the time it took to walk to the house, make the call, and return to the car. I knew Anna had arrived within thirty minutes of the call. We both agreed that the time I had lost was between the stall-out and my arrival at the house. According to Anna, I had called her at 9:15 (it was then 10:00), yet I believed I had called her around 8:15 or 8:30 at the latest.

What had happened? Was the man in the white Stetson something other than he seemed? It would be nearly a year before I discovered more about the incident and how it related to this phenomenon.

14. Anna

Beth tells me I have weirder dreams that anyone she has ever met. She says some of my dreams have no basis in reality, others we can both relate to and determine where the ideas come from. No, they're not dreams of flying saucers or little grey guys, as some abductees report. See, I'm not involved, I keep telling myself. I thought everyone had weird dreams that made no sense—that's what dreams are for, to sort out the feelings and ideas you can't express during waking times. I had a dream the night of January 16, 1992, that for some reason still sticks with me. Most of my dreams are foggy on awakening, or sometimes clear, but by noon they sort of fade and then I just don't remember them anymore. This one I remember.

I was in a swamp. I remember seeing sheep and deer jumping into the water from the right, one at a time and knowing that the creature—an alligator?—and his friends ate these animals in one gulp. An alligator rescued me from the water and put me in a Volkswagen bus and told me that I shouldn't swim there or I would be mistaken for a sheep or deer and be eaten. He then kissed me, put his arms around me and I became calm and everything was okay. Then I woke up or the dream ended. I remembered it very vividly when I awoke, all that day, and still remember it today. Beth and I have talked about other dreams, yet I don't remember them after a day or so. This one is just weird enough to be something.

I used to have a recurrent dream, over a period of several years, that involved driving a car (must have started in high school or college, because I knew how to drive at the time), yet when I'd try to apply the brakes (going down a hill towards water usually) the brakes wouldn't work. Nothing ever happened to me, but it was scary. I assume it dealt with stress—a lack of control in my life. I have a strong need to be able to control my body, my environment and the things that go on around me. It's upsetting to me when I can't. I've often felt that this wasn't the healthiest (mentally) way to live my life, but it works for me. I have a real hard time giving control to someone else, but I've gotten better at it over the years.

Another recurrent dream I have had over the years is of leaping across an open field like a ballerina. I jump so high and then just float down, to leap again. Such freedom! Sometimes the dream changes and I find myself floating along in a city, at about the height of the first story of the buildings. I look down on all the people, but they can't see me. It's such fun to be omnipotent!

I talk in my sleep a lot, and sometimes walk as well. I have been known to get up out of bed, hold conversations for half an hour or so yet be asleep. Yes, that's strange by anyone's standards. My old college roommate, my sister and now Beth, have figured out when I'm really asleep when I seem to be awake. They tell me to go back to bed, and I do. I don't remember any of these incidents.

For the next several months I spent most of my waking hours reading books about UFOs and abductions. I was reading about three or four books a week, and it was affecting my work. I no longer cared about anything I was doing at work; somehow it didn't seem to be nearly as important as what I was beginning to uncover. I read at work, took long breaks so I could read more, went to the library to do research, and kept doing as little as possible that would interfere with my UFO readings. I was lucky that work was also in turmoil. My boss had just left for a new job and several other people on the staff were changing jobs—no one cared much what I did or didn't do. So I read UFO books! I had always been a hard worker before, so I rested on my accumulated knowledge to barely get by, doing just enough to survive. No one even noticed! I don't recommend this to anyone; obsessive-compulsive behavior can lead to more than just mental impediments.

At this point I tried to explain to myself most of the things I had been remembering about my childhood and adult life in terms of the many books I had read over the years. About once every five years I would get on this kick to read self-help psychology books. I knew I was normal, not average, but different than the majority of the people I came in contact with. The self-help readings were triggered by my bouts of depression. The first time was in my first year of undergraduate school, but the psychologist and I decided it was just a phase I was going through—learning to cope on my own in a new world that was not as predictable without the comfort of home and family; not an uncommon reaction to the transition between high school and adulthood.

The second time was when I was living in an island paradise, but for some reason I couldn't cope. The third time spanned two years when my mother was dying of cancer. My work life was miserable, I fought all the time with the men that were in control of my work (what incompetence!) and finally had to be transferred to another job so that I could make a new start and try to become productive again. In each instance, I'd work with a psychologist for a few months, feel more in control, and then quit seeing him. Besides, each time they felt the source of all my problems was a childhood rape, yet I felt that I had integrated that experience and wasn't willing to deal with it again. I knew that wasn't it.

I tried to explain my difference from other people in terms of all those theories I'd found in books. I was an independent woman in a man's society, and my fighting for control of my life was similar to what many of the women of my generation were going through. I have just begun to realize my potential and have so many more choices open to me than my mother ever dreamed of, so I wasn't about to let arbitrary societal rules put me in a one-down position. I was more intelligent and mentally more adept than most of the men I came in contact with. I could have it all!

I read books about middle children (I am the second of four children) and thought there were some interesting parallels to my development. I always felt that I had to work harder than the other children in my family to be noticed, to get the love and attention that I needed. But I told myself it made me a stronger person. I didn't need as much from other people; I wouldn't be dependent on anyone.

Of course I was also a military brat. That life style has been shown to have profound influences on the children of military families. Our lives were rarely settled for more than a few years; we learned to adapt quickly. The moves across country every few years were always exciting for me because of the new adventures to come. But it also left me with a sense of isolation; no longtime friends and an unwillingness to really resolve relationships with people. After all, I could always start again at a new place. Then I had two lives within my family as well. We acted one way when my father was away at sea, but I always felt tense when I knew he was coming home again and I would have to cope with his authoritarian manner. I now realize it was as hard, or harder, for my parents to adjust to these two lives. But that was what was expected, and we all did our duty.

Then there were all the articles on personality type. I definitely have a Type A personality—I drive myself and others ever onward toward new goals and new ways of thinking. The Meyers-Briggs temperament types also helped me realize I wasn't abnormal. My type was INTJ (introverted, intuitive, thinking and judging). It is a relatively rare temperament type (less than 1% of the population), but fully functional. I knew I was special! Now I understood why I never felt really comfortable around so many people until I got to college. More people there have a tendency towards this temperament type. I had found kindred souls and felt I'd finally found my niche. But I couldn't stay in college forever. I needed to be productive.

Now I had all of these reasons for being different from most people I met, but I still didn't have any answers. Each new theory gave me another piece of the puzzle, yet I still haven't sorted it all out. Maybe one never does until one accepts and then begins to feel comfortable with it. Maybe I'll never get there. Maybe it's not important. Life goes on either way.

So where was I about my possible abductions adding to all this pop-psychology? Nowhere. More confused than ever. I still did not consider any of this evidence that I really was being abducted by creatures not of this world.

15. Beth

How were we to deal with this strangeness, this intrusion into our lives? I had been concentrating hard on compartmentalizing each incident, but it was becoming difficult to keep up. I was running out of boxes to put these events into!

We had at least decided that seeing a psychologist might help. We weren't looking for confirmation of beliefs; we weren't even sure what we believed in anymore. What both Anna and I needed was comfort, a calm and reassuring presence who could help us to help ourselves.

But where should we look? Who could we trust to keep an open mind to the *possibility* that what was happening might be real while investigating all the alternatives? This therapist would have to have some knowledge of the phenomenon so that he or she would not be working with us in total ignorance, and therefore, have at least an idea of the stress under which we were living. At the same time, if the therapist was a firm believer in the existence of aliens and the abduction phenomenon, he or she might overlook other hidden psychological problems in favor of those beliefs. This, we both agreed, would be just as bad as one who refused to entertain the possibility at all. So, it boiled down to locating a therapist who was skeptical, but not *too* incredulous. We had our work cut out for us and it would turn out to be a long, hard search.

Finding the right therapist was certainly important, but for Anna, discovering if she might be involved was becoming more important. If these events were real, if these alien beings were not the product of my imagination infecting Anna's creative mind, then there was a good chance she was also involved. Since we met as adults in 1987, I had felt close to Anna, too close for not having met (and possibly known) each other years before, perhaps as children. This feeling became stronger with each passing day and as events unfolded, as more strange happenings occurred, the connection I felt with Anna grew stronger and more familiar. I was almost convinced these experiences were what linked us and maybe what had drawn us together again as adults.

There was more here than either of us could presently identify or explain. Coincidence notwithstanding, Anna and I needed to know what this connection was and how it came to be. And there seemed to be only one way of finding out: hypnosis. Our complex set of rules for choosing a therapist now included finding one who would do regression therapy as part of the program.

This would prove even more frustrating. Not only was the cost of a therapy session exorbitant, meaning that only one of us could afford to go, we were still stuck with the problem of locating a therapist who could be nonjudgmental!

What a mess this was turning out to be! In our desire to find answers, we were inventing more questions.

It seemed prudent to contact the Fund and ask if they knew of anyone who could help us with this dilemma. We were informed that some UFO researchers were qualified hypnotists, but few were licensed therapists. We were given a couple of names of licensed therapists, but of those only a couple had any knowledge of the phenomenon and even fewer would perform hypnotic regressions within the therapy process. We were in a loop with no way out, but we had to keep trying anyway.

So we would feel as if we were doing *something* constructive, Anna and I decided to look for books on self-hypnosis, believing that if we understood the general principles we would be able to help each other and piece together parts of our "missing" pasts. At any rate, it couldn't hurt to try. The least we could expect to accomplish would be to learn good relaxation techniques for helping us cope with the stress we were under.

We eventually found a book on the subject and studied it religiously before attempting to use the technique. Anna had agreed to be the subject while I guided her along. The technique worked well to relax the body and mind, alleviating tensions and presumably allowing the subconscious to take over. It seemed to be going smoothly; Anna was quite relaxed and breathing evenly, so suggesting a date and event we had previously decided to explore (one which had no known anxiety attached to it), I attempted to regress her to that predetermined event. She was able to put herself there in a rather detached state, like an observer, but had difficulty bringing the event to life. It eventually slipping back to the present without our having learned anything we did not already know.

Frustrated by this failure, we decided that although the exercise appeared harmless, I was too inexperienced to know how to guide the subject, ask the right questions, and evoke responses without inadvertently leading Anna and chancing confabulation, a subconscious tendency by the subject to "create" responses that the subject believes the hypnotist wants to hear. Once the subject has confabulated, neither the subject nor the hypnotist is likely to be able to separate confabulation from true memory.

On the first of April we decided to try again; this time I would be the subject and we would explore the event of March 8, hoping to discover what might have happened during the missing time. Having been hypnotized before, I had little trouble dropping into that relaxed state. Fearing that I might become frightened by having to relive the event (especially if something was uncovered), I placed myself in a protected place where I could view the incident as an impartial observer. However, once regressed to that night, I could not maintain the detachment and became a participant instead. Anna guided me like an expert, though, reminding me I was quite safe and had survived the experience.

As I relived the car stalling, pulling off onto the shoulder, seeing the white car pull in behind me with its high beams reflecting in the rearview mirror, and described the man in the huge white Stetson as he approached the driver's side, I started laughing hysterically. The man was naked! I couldn't believe what I was seeing so clearly, as if I were back there again in real life. How could I have forgotten something like that? When the man moved around to look under the hood with his flashlight, I could clearly see he was wearing no clothes, yet I couldn't see any navel or body hair and wondered aloud if he was wearing a very tight flesh-colored body suit of some kind. But that hadn't made any sense, so I accepted that he was probably nude.

When he had moved around to look into the passenger window, I first noticed the thick mustache before realizing he was wearing dark aviator style sunglasses (at night?). I described the oversized head and hat, knowing I had never seen anyone who looked like that, never seen anyone with a head that large. Adding to that the fact that he was nude made the entire image ludicrous. I became frightened, though, when the man seemed to be telling me something without moving his mouth. I cringed against the driver's side door, screaming for him to get away. Anna, realizing I was close to panic, brought me back out of the hypnosis state quickly, apologizing for letting things get out of hand.

Obviously we were toying with something that could quickly get out of control and may cause us real harm, despite our attempts to keep things under control. There was much to discover here, much repressed information which might tell us something about what was happening, but we were treading on dangerous ground. As much as we wanted—and needed—to understand, the wisest thing to do was to leave regression to the experts, no matter how long it took.

16. Anna

I was no closer to solving the puzzle of Beth's (and perhaps my own) involvement in UFO abductions despite all my extensive reading and theorizing. We needed more help than we were getting, yet I was unwilling to commit large sums of money, which I didn't have, to therapists who had little knowledge of abductions. Our area seemed particularly devoid of knowledgeable therapists and hypnotists.

Beth had mentioned that many years ago a friend of hers was a hypnotist and he had taught her self-hypnosis. She found it relaxing and restful. So, off we went on our weekly trek to the used book store to find some books on self-hypnosis. I, of course, bought a technical book on hypnosis, and read it thoroughly. Boy, was it boring. I finally read the book that Beth had bought and found practical applications of self-hypnosis. I knew we weren't going to use hypnosis for stopping smoking or dieting, but for a much more useful purpose: uncovering memories. I was scared. I was on the brink of finding out if I really was being abducted by aliens. But did I really want to know? On some days, yes.

I read the book several times before trying hypnosis. I found I could relax, could make my fingers and toes tingle on command, yet I couldn't retrieve memories of any kind. Oh, well. I guess I just needed more practice. But practice didn't help.

One evening Beth and I decided to try to help each other with hypnosis. She'd been having the same kind of results that I'd had. She described to me how to mentally create a safe place to start and end the hypnosis session. I chose a place I'd always felt safe and comfortable—under a pier submerged in about thirty-five feet of seawater. Not such an odd choice for a person who lived to scuba dive for ten years of her life. Beth helped me to become hypnotized and I felt very relaxed as I imagined myself within the confines of the pylons where nothing could get to me. It was a great place to be. When she asked me to go back to a definite time, one where we were reasonably sure that nothing had happened connected with aliens, all I could see was blackness. No images came. I thought I was supposed to see things! A frustrating experience for me. I felt sure we had failed because I just couldn't be hypnotized, despite what the books said, and I would never be able to find out if I'd ever been abducted. I almost cried. Now I would have to rely on only my conscious memory, and that was notoriously poor in this area. It hadn't helped me so far.

One other evening, we decided to see if Beth could uncover some memories of the incident with the guy in the white Stetson. I helped Beth go to her safe spot, a large room where she held the only key, and then asked her to go back to March 8. She did it! I was amazed, and jealous. Why was it so easy for her?

It was fun asking her questions and getting her to respond. Although I did have to prod her to talk to me, instead of just seeing the story unfold in her mind. This made me feel like a magician, in control. But what a burden. I had to keep figuring out where in the story she was, yet without leading her to make any assumptions about what was happening based on my question. I had read several transcripts of other abductees' regression sessions and knew some of the questions to ask, and others not to ask. I also felt responsible for her well-being, and I had to get her back safely. Was it possible to stay hypnotized forever? What happened if she didn't come out of it? My mind was full of doubts.

When Beth started laughing hysterically, I really got worried. But when I finally convinced her to tell me what was so funny, I also broke up. A nude man with a huge hat! I guided her through the next few memories and when she started crying in fear, none of my reassurances made any difference. I decided we'd had enough. I tried to bring her out of hypnosis quickly, but she beat me to it.

We talked afterwards and decided that we wouldn't ever do that again. I was more scared than she was, and felt responsible for her panic. She knew she could end the session any time she wanted to, but I didn't. We agreed that from then on we'd rely on conscious memories or an expert hypnotist.

chapter 5

Awareness

17. Beth

Facing my father across the dining room table, I almost changed my mind. I had thought it would be easier than this. *"Just spit it out,"* I said to myself. *What's the worst that could happen?*

I had so many questions about my childhood, so many odd memories of events that, as I grew older, left me feeling confused. Had the strange occurrences I remembered actually happened or were they just typical of most childhood memories? As we grow up, most of us recall only those things that were common to other children our age, things which were not necessarily traumatic or unsettling. I suppose it's a defense mechanism that's quite natural. But when our adult world is bombarded by eerie happenings, sometimes childhood memories are stirred, memories long ago repressed. That was what I now suspected; somehow these current experiences (real or imagined) had awakened memories of similar events still left unexplained from my childhood. By talking with my father, I hoped to clarify some of these disturbing glimpses, these brief flashbacks that continued to plague me.

So I gritted my teeth, looked my father in the eye and asked him if he remembered my having experienced anything in my early years still left unexplained today. He sat quietly, appearing to be thinking back and trying to remember. Then, quite unexpectedly, he threw the question back to me: What did I remember? I had planned to cover several events, beginning at about age four and on through my early teens, but I had not planned on itemizing them *first*. Since it was apparent my father did not wish to influence me, I chose a rather innocuous event that I only recalled in disjointed pieces.

I was around four or five and had awakened during the night feeling very frightened, but not knowing what had frightened me. I recalled looking out the bedroom windows which were on the far side of the room next to my sister's bed. There were "cats" staring in through the windows—or what I thought were cats—because their eyes were slanted and had vertical slits. (See Figure 1.) I tried to wake up my sister, but she either didn't hear me or I wasn't calling her as loudly as I thought. I screamed for my father to come, but don't know if he ever did. I couldn't remember what had happened after that.

Sister, age 6, would not wake up.

Seeing the "cats" through the window at age 4.

Figure 1. One of Beth's early childhood memories, as confirmed by her father.

I do know I often awoke in the mornings hiding in the bedroom closet, the clothes pulled from the hangers and draped over me, the closet door closed tightly with the cracks stuffed with whatever clothing would fill the gaps. I had no idea why I did this. On other occasions my parents would find me huddled against the inside of the bedroom door making it difficult for them to open the door until I had moved.

My father had listened patiently through this first recollection, making no comments other than an occasional nod of recognition. When asked if these events had actually happened as I remembered, he took a deep breath before answering.

"You hid in your closet a lot when you were little. I remember having to dig you out of the clothes pile. You even had the cracks in the closet door stuffed, just as you said. When you weren't in the closet, you were blocking the bedroom door by lying in front of it."

"But why would I do those things?" I asked him, almost afraid of the answer because somewhere on a very deep level I already knew the reason.

He shrugged. "I believe you were unreasonably afraid of light. Maybe your eyes were simply sensitive to light. I don't know."

Light? But why would a little kid be afraid of light? Most children were afraid of the dark! Then I realized I had always had an unnatural fear of bright light. I can't tolerate overhead lighting and have trouble going to sleep unless my bedroom is completely dark. Any outside light, from the moon or stars, is effectively blocked by heavy drapes and shades. Light escaping from the hallway is muted by stuffing pillows or clothing in the cracks of the bedroom door—just as I had with the closet door when I was a child. But what did this mean? How did it relate to what was happening now? Was it, instead, a symptom of some psychological problem which was never explored?

I asked if he recalled my ever complaining about cats peering in the windows at night. Had I ever called out to him during the night about the cats? He admitted hesitantly that I did sometimes scream during the night, but that I usually went right back to sleep, adding that I had always liked cats. "Don't you remember your stuffed cat collection?"

I did. But it wasn't cats I liked, it was kittens. We always had cats when I was growing up, but I ignored them once they were no longer cute little kittens. I did remember begging for one of those furry little "bed kittens" that were so popular with little girls at the time, insisting that was the only thing I wanted for Christmas. I was around five when my father finally found one of those kittens, a white ball of fur with a bushy tail and a pink bow around its neck. I named it Dee-Dee Kitten, though I didn't know why I chose that particular name for the stuffed animal.

I kept the kitten on my bed in front of my pillow and instructed it to keep me safe. From that time until my early teens I wanted nothing else for birthdays, Christmas, or other special occasions than another one of the stuffed kittens. I named them all Dee-Dee Kitten, though I considered them each individuals and could easily tell one from the other. By the time I was thirteen my father could no longer find these stuffed animals and I had to accept that my obsession with collecting them must stop.

(Note: It was during these early teen years that I became friends with a girl named Dee. She had also been enamored with the little stuffed bed kittens and had owned several by the time we met. Dee and I had a number of shared experiences during our high school years.)

This memory stimulated another: I had chosen certain Dee-Dee Kittens to accompany me outside (some I only permitted in other rooms within the house while others I restricted to the bedroom), and often one of these "outside" kittens were with me when I would find myself out of the house late at night, unsure where I was or how I got there, but knowing my father would soon be along to collect me and take me home. When I mentioned remembering incidents of being out late at night while still very young, my father laughed without humor.

"You were gone more than you were home!"

I was startled by this statement, especially since I considered my father to be very protective of me as a child and couldn't imagine he would have allowed me to wander around outside late at night! What did he mean?

He suddenly looked stricken, and I swallowed the urge to question him further about this unsettling memory. We sat in silence for several minutes before my father finally spoke, his voice raspy with emotion. He sensed there was more to this stroll down the memory lane of my childhood and asked that I please tell him what had prompted these questions. I had hoped it wouldn't come to this, hoped he would explain all these incidents away as the distorted memories of a normal child with an active imagination. In that way, I would not have to reveal the strange events I had experienced of late and how I felt they might be connected to the disjointed memories of childhood.

I straightened in my chair and looked away from him, unsure how to begin, but knowing there was no hope of concealing it any longer. I began with the December 1991 incident, relating all I could then remember of what had occurred, and continuing uninterrupted through the most recent experience, looking down at my folded hands and feeling foolish. It really did sound crazy. I wouldn't have believed it if someone else were telling me these same stories.

18. Anna

As Beth began to explore childhood memories with her father, she naturally asked me if I had similar tales to tell. Of course not. I said, "No, my family is normal." The more we talked, the more I realized I would have to think back on my relationship with my parents and siblings. But not right then.

Beth visited her parents nearly every Saturday evening after chores and did not return home until Sunday evening. I found myself envious of her seeming freedom to discuss this with her father, yet worried that something strange would happen to her on the journey to and from their house nearly sixty miles away. After all, odd things only seemed to happen after dark. The trauma of her involvement with the alien beings had strengthened our friendship, even though we hadn't thought that was possible. We were already closer than most sisters, but that closeness had intensified to a desperate need for emotional support. Beth, seemingly, needed it more than me.

Occasionally, one of us would go out of town on business, and I found myself calling home as soon as I got to the hotel, or anxiously waiting at home to hear from Beth that she had arrived without incident. I dreaded making or receiving the calls, expecting something to have happened to Beth in my absence, yet I needed to know.

Once, when Beth attended a weekend seminar and stayed in a small bed and breakfast at the southern end of the state, she didn't call until I had waited anxiously for several hours. I knew something bad had happened to her. When she finally called, she had another strange story to tell. Not bad, just strange. She'd had no interference from anyone, or anything, on the daylight drive down. She just hadn't been able to call me when she got there. Seems that as soon as she started to call home, the phone went dead in her hand. She waited for a few minutes, then tried again. No luck.

When she walked downstairs to check with the owners, she found that all eighteen phones in the bed and breakfast were out of order. This had never happened before and they apologized profusely, saying that they would call her as soon as service was restored. It took several hours to get a dial tone, even though the telephone company couldn't find anything to fix. We laughed about it and decided it was one of the vagaries one might expect on expeditions to small rural towns.

We wouldn't have thought any more about it, but for an incident about three weeks later. I went to Norwalk, Connecticut on business, stayed in a large hotel and the same thing happened! When I tried to call home, the telephone in my room lost it's dial tone while I was dialing. It took only about an hour to get telephone service restored to my hotel room, so Beth hadn't panicked by the time I got through to her. Interference by aliens, or someone else? Coincidence? Probably, but we were beginning to get paranoid about almost everything that happened that didn't fit our old, established patterns and beliefs. We had been assaulted by too many strange things in the last few months so that our concepts of what was possible, real or extraordinary were rapidly changing. We clung to each other for emotional support and reality checks. We were both good at playing the skeptic. One of us could usually remain detached and logical when the other was being traumatized by some real or imagined event.

I found the constant checking in, phoning home every night while I was away, phoning from work a couple of times a week when I felt uneasy or thought I'd had a rough night almost annoying. I hadn't checked in with anyone since I left my parents' home, and often rebelled against it then. I am fiercely independent and expect others to be self-sufficient. But I was worried about Beth, a new emotional response for me. I needed time away, to immerse myself for a week in other people's boring realities—to have nothing new come my way, no double meanings in phrases, no strange events lurking to entrap me. No hint that everything was not as it seemed. I still called Beth each evening I was gone, but I prayed (not literally, I gave up organized religion a long time ago) that everything would be quiet until I got home. It usually was.

I was still unwilling to examine my family relationships. It was too hard. I knew by now that alien abductions sometimes ran in families, but since I wasn't involved, my parents and siblings couldn't be either. Beth's memories of being terrified by "cats" at her bedroom window at night and her fear of light in her bedroom had never happened to me. In fact, I preferred to sleep with the lights on. Ever since my mother read an article about winter depression being exacerbated by insufficient exposure to sunlight, I'd used it as a good reason to sleep with some light in my bedroom. Of course, I had to switch to a nightlight when Beth mentioned that my light, reflected in the hallway, bothered her, but it was enough. I could still see if I woke up in the middle of the night. If light could actually lessen my depressive spells, so much the better. Depression isn't much fun—for me or anyone around me. I don't recommend it as a coping mechanism for anything.

Beth and I had adapted so much in the last few months, I was sure that soon we would be able to sort this out and return to what other people considered a normal life.

19. Beth

Wading through the progression of strange experiences and dis-jointed memories, having to relive each one as I related it to my fa-ther, left me drained and embarrassed. Would he believe me? Would it be better if he didn't?

It was only then that I realized my mother, who had been sitting in the living room reading, was no longer there. Concerned, I asked my father if she was upset by our conversation, hoping she hadn't misinterpreted my probings as accusations. In response, he only shrugged, as if to say her leaving had nothing to do with what we were discussing. On a deeper level I felt it had a great deal to do with it and wondered how I was going to approach my mother with sim-ilar questions about her remembered odd behavior.

Setting that challenge aside for a later time, I told my father about a disturbing memory flashback I'd had on April 4.

I was only fourteen when I became friends with a nineteen year-old boy named Eddie. This relationship, though innocent enough, proved traumatic for both myself and my family. I had kept the friendship secret, understanding that my parents would not approve of it, no matter how platonic.

I was very young, but not very naive. I knew how women be-came pregnant, understood that kissing—no matter how passion-ate—could not result in pregnancy. Yet two months after meeting Eddie, I was somehow very much pregnant. At first I ignored the symptoms; morning sickness, cramping, breast tenderness, not un-derstanding that these were the classic signs. When I finally com-plained to my parents about these problems, they also looked for other causes. Eventually, as the symptoms continued unabated, my father took me to see our family doctor for an examination. The doc-tor, not so inclined to ignore the obvious, took urine and blood sam-ples before sending me home for bed rest until the results were back from the lab. I stayed home from school for two days, worried that I would miss final exams. The school year was almost over—only a month left before summer vacation—and I did not look forward to having to attend summer school to catch up!

True to his word, the doctor called, asking that my father bring me back into the office at his earliest convenience. My father took the next day off work to keep the appointment, apparently believing that I had some serious illness. Upon arrival I was left in the waiting room while my father was escorted into the doctor's office for private con-sultation. When they emerged from the office, their expressions told me only one thing: I was not dying from some virulent disease, but my life was nonetheless hanging by a thread. They were angry, my father in particular, glaring down at me as if I might have committed the most horrible of crimes.

It was not long after that I discovered why my father and our trusted family doctor were so upset. *The rabbit died.* Three little words I never expected. Why did the rabbit die? I asked them. What did that have to do with what was wrong with me? I didn't know anything about rabbits!

I was not left in the dark (with the dead rabbit) for long. I couldn't remember what the doctor had told me during that tense visit, and very little about my later conversation with my mother and father after returning home to face the music, but I had clear memories of complete denial. I could not have been pregnant. It was not possible. I had not had intercourse with anyone, I proclaimed adamantly. Why wouldn't they believe me?

(I, of course, know now why they couldn't believe me. I would not have believed it of anyone else even then.)

I was immediately grounded, any more denials futile. I was driven to school by my mother in order to take final exams, then picked up from school afterward and taken directly home. I was not permitted to receive phone calls nor was I allowed out of the house for any reason. Confused and angry, I pouted in my room and prayed that my Eddie would call so that I would be exonerated. But he didn't call. Somehow I managed to get through the school year without failing my exams, but I was still on full restriction, cut off from my friends without explanation. They soon gave up trying to call and I was completely alone with my burden, still not understanding how I could have gotten pregnant.

By July my parents had come to the decision that I would have to either be tutored the following school year or sent to a private school until the child was born. I was consulted about this and told that I could make the final decision. I felt they wanted me to stay close, and even though they were obviously disappointed in me, they were not ashamed of me and encouraged me to take good care of myself and not worry about what would happen after the baby's arrival. That could be decided when the time came—and that decision would also be mine to make.

The month crept along while I struggled to choose between tutoring and private schools. I remained aloof from friends and relatives even after restrictions were eased.

Then one night in mid-July, after I'd gone to bed, I knew Eddie was waiting for me. He was at the high school, some two miles from my home, and wanted me to meet him there at 11:00 P.M. that night. He hadn't called, hadn't come by, but I heard him nonetheless and knew I had to go. I got up and dressed, then slipped back into bed and pulled the covers over me. It was still early and my parents had not yet gone to bed so I would have to wait. I was excited by the prospect of seeing Eddie again and it didn't seem possible that I could have dropped off to sleep!

I awoke with a start and saw it was almost 1:00 A.M. Jumping out of bed, I listened for any sounds from the house. It was very late and surely my parents and sister were sound asleep. I scrambled onto the dresser and climbed out the bedroom window as quietly as possible, worried that Eddie might have given up on me by then. In my rush to get out the window I tore my pants, but didn't slow down. I dropped into the shrubbery below, scratching my face and hands, jumped to my feet and raced the two miles to the high school, dodging car lights by ducking behind trees and bushes. I couldn't afford to be seen. It was too important that I get there before it was too late!

I dashed through the school parking lot, which was fortunately behind the school and therefore not visible from the main road, and headed for the bleachers lining the football field. Hiding in the shadows, I stopped to catch my breath, praying that Eddie was still there. I waited for what seemed to be hours, though it was probably no more than a few minutes. At last I saw a vehicle pull up in front of the school and stop, it's headlights shining like beacons in the darkness of the deserted school lot. It looked like an ordinary school bus. It was not what I expected to see.

I waited another agonizing minute, afraid the bus was a trap of some kind. I called Eddie's name, but there was no reply. Well, I thought, I might as well go see since I was already in pretty deep and if I went home again it would have all been for nothing and I might never see Eddie again. Steeling myself, I rose from the shadows and walked out to the bus. The door was open so I climbed in and sat down. There was no one else on board and the driver (if there was one) did not acknowledge me. The bus door closed and we pulled slowly away from the curb.

Here my memories came to an abrupt halt. My next memory of that same event, though much later, apparently relate to the experience's conclusion. I recalled waking from a sound sleep and discovering I was on a Greyhound bus entering some strange city, a city I'd never been in before. An older black woman, well dressed with white gloves and flowered hat, was seated next to me and looking at me with concern. She asked if I had run away from home. I didn't know where I was or how I'd gotten there and didn't think I'd run away from home, but I nodded anyway. She asked if I wanted her to call someone for me, to have someone pick me up and take me home, but I shook my head this time. Finally I said I'd call my parents. I asked the woman where we were and she said, "Why child, we're in Little Rock."

I didn't know where Little Rock was and said so.

"Arkansas," she said, frowning at me. "Little Rock is in Arkansas. Didn't you know that?"

I said I did, I'd just forgotten because I was only half awake. She offered her help again and I thanked her, but said I'd be fine. Just then the bus pulled over to the curb and the door opened. It was the end of the line and everyone was getting off, so I followed the half-dozen passengers out. The black woman pointed to a restaurant in front of the bus stop, explaining that it would open for breakfast in a few minutes if I wanted to wait. It was still early; the clock over the building across the street read 6:15. They had a phone inside, she added, then she walked away without looking back. Seeing a bench a few yards away, I went to it and flopped down, wondering if I was dreaming all this and how long it would be before I woke up.

I saw a familiar car drive down the quiet street toward me and pull up to the curb directly in front of the bench. It was my father. I had never been so glad to see him. He leaned across the seat and opened the passenger door and I jumped in. Neither of us said a word to each other. When we had cleared the outskirts of Little Rock, we stopped in a small diner and ate breakfast in silence. We arrived back home late that same night and I went directly to bed, neither of us having said so much as one word to each other during the entire journey. My mother came into my room and kissed me goodnight, but also said nothing about my unexplained absence.

The episode was never discussed, recriminations never came, though I was constantly prepared for them. I felt wonderful despite the eerie time loss, better than I'd felt in some time. The morning sickness seemed to have subsided and my breasts, which had been so tender I was unable to wear a bra, seemed to shrink and I felt almost normal again—except for having started my period. Perhaps I had dreamt the whole adventure. I decided I had.

The following day was my scheduled prenatal exam with our family doctor and I found myself, to my surprise, completely at ease, assuming I had come to accept this strange pregnancy no matter how the condition had developed. The exam took much longer than expected, punctuated at regular intervals by the doctor's "humphs" and "huhs" with no translations forthcoming. At last I was asked to dress and come into his office, where when I arrived I saw my father already seated. I sat next to him and waited for the doctor to speak.

"You're no longer pregnant, Elizabeth," he announced without preamble.

I blinked, but didn't know what I should say. Had the rabbit come back to life?

My father looked stunned into silence. The room was so still I could hear my heart beat. Finally my father found his voice and questioned the doctor about an ominous sounding thing called a "miscarriage." The doctor said he'd found no evidence of a miscarriage, no scar tissue, no excessive bleeding, but that I had indeed begun a normal menstrual cycle and that overall I appeared perfectly healthy and definitely not pregnant.

We left the doctor's office in high spirits and life went back to normal. The strange pregnancy was never mentioned again.

After so many years of silence on my "strange" pregnancy, my father was understandably reticent to discuss it now. I encouraged him to be honest with me and to feel free to tell me whatever he remembered of that time, even if what I recalled had never happened. It would, in fact, have been a great relief to hear that this painful memory was totally false. My father did not oblige, however, and confirmed my worst fears: It had happened. I had been pregnant. He had driven to Little Rock to pick me up and take me home. I was not pregnant when I next saw the doctor.

"But how did you know where I was?" I asked him suspiciously. "I didn't even know where I was!"

He couldn't explain, only said he always knew when it was time and where to find me. He had never questioned this knowledge, but had often tried to resist it, unsuccessfully.

Did he know why this happened? I asked.

No, he admitted. He just knew he had to.

I was furious with him! How could he sit idly by while his young daughter was out traipsing the countryside in the middle of the night doing God knows what with God knows who? If he was aware I was gone and knew where to pick me up, why couldn't he stop me from going in the first place!

"How could you let this happen?" I yelled. "You're my father! You were supposed to protect me!"

I saw the tears forming in his eyes and felt immediately ashamed of myself. I'd never seen my father cry and I didn't want to see it now. I was not handling this well. This was not a time for recriminations; this was a time for understanding.

He gathered himself together and tried to explain that he had no more control over what happened to me than I did. The only protection he could offer was to make sure I got home again safe and sound. He could not, he assured me, stop it from happening.

All at once I understood. My father had listened to these horror stories in reasonable calm because he knew the stories already. He knew what was happening to me—even now. He just hadn't known before that I remembered any of it.

I then told my father about the strange beings in my flashbacks, but found it difficult: There were no words to adequately describe them and nothing in my experience (I thought) with which to compare them. I had made a few sketches from my brief memories, but was hesitant to show them to him. I didn't want to frighten him, but I didn't want him to think I was crazy either! Putting up a brave front, I withdrew some of these drawings from their hiding place in my overnight bag and spread them out on the table. Two of them were of my father and me standing before one of these slender creatures; I recalled being about four or five at the time and represented myself as I might have looked then.

He examined each one carefully, nodded, then looked away. I asked him if he remembered us ever being in the presence of beings that looked like the ones I'd drawn. He turned back to me, slowly, and sighed. Yes, he admitted, but he'd always thought they were nothing but nightmares.

"I think they're real," I ventured, "and you've seen them too, haven't you?"

Yes, he had seen them. And he had seen them before I was even born.

20. Anna

Parents. Family. What mixed feelings we carry into adulthood. On the surface, everything was normal. Sibling jealousies, disappointments, family outings, friendships and learning how to live in an adult world; it was all part of growing up. If I really think deeper, I find long-held resentments about how I thought my parents always treated the other kids better than me. There was my older sister, the first born, the apple of my dad's eye. Then me, the only blonde in the family. My younger brother was my mother's favorite kid. He always seemed to need protecting. My other sister was the youngest, and the baby of the family always gets spoiled. I always had to be better than anyone else at everything I did so that I could meet my father's standards, and so that my parents would love me more. Pretty normal stuff.

It's not that my parents weren't supportive; in some ways I think my father was too supportive. I now appreciate his willingness to teach me everything he knew and to always tell me that I could achieve anything I wanted to, if I only worked hard enough. I had no limits on what I could achieve. I only hated him for a few weeks, when I was about twenty-seven, when I realized that what he told me wasn't true in the world I lived in. There were many situations that would interfere, and many people who would stop me from achieving my goals. Intelligence and hard work would never be enough. I had no political skills, I was always up-front and honest with people, I had never learned to play games. I had no tact; my world was very black and white, good and bad, right and wrong. When I finally learned to see things in shades of gray (no pun intended), I was at another stage of adulthood. It was a hard, but necessary, adjustment.

My mother grew up in a completely different world than her daughters. She married young, at nineteen, and had to drop out of nursing school when it was discovered she was married. But she was happy with the nomadic Navy life after growing up in a small town. She had four children in nine years, and seemed content with domestic life. She did get a job as a seamstress once, when we were all in school, but soon quit because my father expected her to be home when we came home from school and to have dinner on the table when he got home.

She taught me all the domestic things that girls were supposed to know—cooking, sewing, knitting, cleaning. She was an accomplished seamstress and made all of our clothes until we got into high school and thought we should have store-bought things. Luckily we had more money by then and could afford to buy more clothes. She was a gourmet cook and taught me how to cook with extravagance as well as inventiveness. To this day I am a master at taking leftovers from the refrigerator and turning them into something completely different and delicious. We spent many years in comfortable companionship working together in the kitchen. The cleaning habits never stuck, though. I'm basically a slob. Yes, I have standards for cleanliness—they just happen to be very low. I can live comfortably with dust bunnies under all the beds and a sink full of dishes.

As I got older I realized there was so much about my mother I never knew, or knew how to ask about. And now she's dead and I don't have that opportunity. After I was in junior high school I always felt I had to protect her. From what, I didn't really know. Society was changing so fast, so many new ideas and opportunities were opened to me that she never had the chance to explore.

I remember one day when I was home from college; we were cooking in the kitchen and talking about my boyfriends. I was shocked to learn that my father had been her only boyfriend and that she had never even thought of having an affair with anyone. I was on my umpteenth boyfriend by then and didn't dare tell her how many I had slept with. Somehow I didn't think she'd understand. Other girls may sleep with boys, but not her daughter! I think now she probably knew, but I felt I had to protect her from that knowledge.

I somehow also felt I had to protect her from my father's ire, and his domineering presence. He always knew what was right, and how his world, and hers, should be ordered. I don't think I ever managed to help much in that area. But I did offer her friendship and love. Even if I felt it was never enough to make up for all the things she had missed out on by being born in the wrong decade.

Obviously, I felt I hadn't done enough when I discovered her alcoholism one Christmas. I knew she had been behaving irrationally, alienating their few friends with her sarcasm, but I hadn't spent more than a week at home in years. I did what I thought was right at the time; I told my father. What a hell I made her life for the next couple of years. But was it really any worse than before? I doubt it. My father took it as a personal insult and forced her to a doctor, put her on Atabuse, and watched as she took the pills every morning.

Eventually they were able to work out some of their differences. She became sober for the rest of her life, and I think he even forgave her for being weak and letting him down. I spent some time at Al-Anon to try to get rid of the guilt I felt for turning her in. Funny, none of us ever talked about the roots of her alcoholism. What triggered it? What was it she couldn't face that led her to make alcohol her coping mechanism?

The only phobia I remember her having was about those paintings of large-eyed children that were so popular in the sixties. She didn't just dislike them, she hated them. There was a real fear in her eyes as she looked at them. I just dismissed it as an overreaction. Could this be related to seeing the gray shits? She never had any missing time that I was aware of, although on a couple of occasions we would find her passed out in the bathroom or the laundry room. We attributed this to her alcoholism. She was alone in the house most of her life, with us kids at school and Dad at work, yet had few outside interests. Why didn't I ever wonder what she did all day? I assumed that cooking, cleaning, gardening and sewing took all her time. It probably did.

The only other odd thing in my mother's family is her great-aunt. Auntie has been found wandering around in a nearby town, or at the farm, in a dazed state, not remembering the last few hours or how she got there. Doctors have diagnosed it as Alzheimer's disease. Maybe it is.

I remember one incident when I was terribly confused as a child. I must have been about seven years old. My cousins and I were outside playing in the snow and had taken my brother along on a sled. We were gone for several hours and when we got home, my brother's boot was missing. We had no idea what had happened to it. Our mothers were furious: How could we have been so negligent as to lose his boot and not even know it? Luckily, his foot was not frostbitten, so the incident was soon forgotten.

The main traumatic incident in my childhood was the rape. It has affected my relationship with men throughout my life, but not that drastically. At times I really hate men. Not all men, mind you. Just the generic classification of man. There are days when I have no use for them except for sex and opening jars. Why were they ever invented? Why can't they go live in their own violent, competitive, power-hungry territory, and leave us women in peace? We'll call when we need one. But, most days, I enjoy being around them. It makes me feel good to cooperate and even outwit them. Such a contradiction in feelings.

But, back to the rape. I had come to terms with it, and had relegated that unfortunate incident to one of the storage boxes in my head. My father raped me. When I was young we used to go fishing all the time. He loved it and so did I. We'd go as a family on weekends; sometimes my father would take my brother and me out on a weekend evening to catch fish. Sometimes my father and I would go alone. It was on one of those evenings when we were fishing alone on the canal that it happened. I was twelve at the time. I don't remember much about it except that he told me it was for my own good, and that he loved me. I don't remember any violence, I don't know how my pants got off. When he entered me, it hurt, I cried, he stopped. Very strange rape. We never talked about it. I never told my mother or anyone else about it. I knew something had been done to me that I shouldn't ever talk about. It never happened again.

After this incident I was afraid to go fishing with him alone. I remember going fishing with my family one evening soon afterwards and being terrified of the land crabs around the canal. I refused to get out of the car. The crabs were migrating and there were thousands of them, but they had never frightened me before. I have a great interest in and curiosity about all living things, and had no fear of animals. I rationalized that they were the excuse to not go fishing anymore. My mother tried to cajole me to get out of the car, but I wouldn't budge. I must have buried that fear because I went fishing with my family, and with just my father (I think), many times after that until I left to go to college.

My father's family never mentioned anything odd that happened to them. No strange relatives lurking in the wings, no family secrets, just an average poor mid-western family that did the best they could for their children. I didn't even resent (for very long) his mother's comment when I was born, "They must have switched babies in the hospital! There has never been a blue-eyed, blonde Jamerson in this family!" They finally accepted that I was my mother's child (and my father's) as I matured to resemble my siblings, even though their first grandchild (my brown-haired, brown-eyed sister) always received the more expensive clothes, birthday and Christmas presents. My mother helped me understand their actions, and made doubly sure I was loved by her. I have always felt closer to my mother's family.

No, I couldn't find anything in my family tree to indicate that we were involved in alien abductions. I could now concentrate on helping Beth search through her memories and experiences, and in the meantime continue my quest for new knowledge. I really was going to get to the bottom of this and find out why the aliens were abducting people.

21. Beth

This stuff was going to scare the bejeesus out of Anna! I was frightened, too, but still I could hardly wait to get home and tell her what my father and I had discovered.

Had my father had these experiences when he was a child? Was I somehow connected through him? Was my mother involved as well? What about my son—and my granddaughter?! My father had hinted that *his* father may have had some strange experiences as well, and his brother, too.

One vividly recalled event occurred near the family's summer home on the Virginia shore. My father and his brother were scanning the beach for colorful shells and stones, each competing to collect the most specimens before being called in for lunch. My father recalled stooping down to retrieve a particularly colorful shell. When he'd stood up again to show his brother, he was gone!

A dense mist had rolled in to shore, blocking out the bright morning sunshine that had been there only seconds before. He called out for his brother. There was no answer, so he began walking along the water's edge looking for him. A shiny object in the sand caught his eye and he stopped to examine it. When he looked up again, his brother was back, the haze mysteriously gone and the bright afternoon sunshine reflecting off the white sand.

His brother was as startled as he, demanding to know where my father had been! As they compared notes, both began to realize something very strange had happened. But what? Should they tell someone about it? Still undecided, they made their way back up the beach to the house where lunch was surely waiting. When they arrived, my grandmother, apparently near panic, scolded them for frightening her half to death. There were two policemen inside the house, my grandmother explained gravely. She had reported the boys missing when they hadn't returned home by 3:00 that afternoon, fearing they had gone too far out into the surf and drowned. Didn't they know people were out there still looking for them? The errant boys were ushered into the house, then confined to their room, where they spent the remainder of the day trying to figure out what had happened to them. Neither had been able to figure out how it had gotten to be so late. One moment it was late morning, about 10:30, the next it was late afternoon—the hall clock read 3:45.

Other unexplained incidents occurred during my father's adolescent and teen years, then suddenly ceased until his twenties when he experienced a number of missing time episodes. Normalcy returned during the first few years of my parents' marriage and after my sister was born, but by the time I was about three years old, my father was once again embroiled in these mysterious events—this time with his young daughter in tow.

More recently, when my father and three of his friends were on a hunting trip in Colorado, all four of the men were mystified when they couldn't account for more than six hours. They had gone into the mountains at sunup and stationed themselves in deer stands perched in trees surrounding an open meadow. Herds of antelope frequented this high meadow in the early morning hours so the men were confident of being able to take at least one good buck between them. From their vantage point high in the trees, virtually invisible from the ground, the men waited patiently for the herds to arrive.

Then, suddenly, it was mid-afternoon! None of them remembered falling asleep or daydreaming the morning away, and were understandably flustered by their inability to account for several hours of prime hunting time. Bewildered and annoyed, they made their way back down the mountain empty-handed.

My father has not gone hunting with these same men since, and they have, he feels, gone out of their way to avoid him.

Retelling these horror stories to Anna, I suddenly realized my denial structures were breaking down. Could these eerie tales be the result of my father's active childhood imagination? Possibly. But what about his brother, my uncle? Would he support them? And what about my childhood memories which my father had so reluctantly confirmed? Surely he would not have falsely done so knowing the pain they had caused me!

Every time I came up with an alternative explanation—imagination, faulty memory, deliberate (yet unreasonable) lying—this other, even less acceptable interpretation superimposed itself. Could this abduction stuff really be happening? Could it have been going on all my life, all my father's life, yet not come to the forefront until now?

Anna speculated that perhaps the memories had been repressed because they were so traumatic. Did this mean they were less traumatic now? I didn't think so. And if they were so traumatic that the mind defended itself by forgetting, why had my father not repressed his memories?

"Maybe he did," Anna offered. "You're just now getting some of your memories back, questioning your father about events from your childhood that don't make sense to you. Maybe your father began remembering weird stuff about his childhood when he was about your age, too."

I admitted that could be the case, but wondered aloud about the drawings and the fact that my father seemed to recognize them. "Why didn't he question them, or say they were only a product of my imagination?"

Anna didn't know, but suddenly I had a revelation. The drawings were mostly of my father and me (as a child), both of us in the presence of these strange looking creatures. "I think I know," I said. "My father remembered these experiences with me as a child, but since I didn't seem to remember them, he kept it to himself. I believe he never expected me to remember, either!"

"You must have given him quite a shock!" Anna laughed.

Not nearly as much of a shock as he'd given me. And it would take a lot of internal probing to work through my emotions, feelings of resentment, abandonment and fear that these experiences have generated. Could I do it? I didn't know.

I'd always considered myself a strong person, able to handle most situations calmly and rationally—without emotion. But this was different. These situations had been occurring for years, and because of the emotional context of the events, I had protected my sanity by submerging the memories. Was I really strong enough now to dredge them up and deal with them in the real world?

Was this the real world?

22. Anna

Beth came home from her weekend with her parents full of news. She was both excited and frightened by the new information. I was thrilled. I wasn't emotionally involved in these past experiences and could look at them in light of my recent readings. New pieces of the puzzle were falling into place.

They had talked about when she was pregnant, when she was fourteen. He confirmed it. Asked how he knew she was in Little Rock, he told her that this was not unusual. Seems that from 1954 to 1962 she disappeared regularly, and he would always know when to go, where to pick her up and bring her home. They never talked about these incidents. I had the feeling, while talking to Beth, that she had disappeared for only a few hours each time, except for Little Rock.

She was upset with her father. Why didn't he ever tell her about the incidents? I tried to reason with her and tell her that he was probably as blocked by the aliens as she had been. They were both told not to talk about it, and her not to remember. It didn't help much. She still felt he could have been a better father and told her what was going on. She wasn't too rational that night, yet I couldn't blame her.

Beth had an abduction dream. I didn't know what to make of it since I was in it with her. We were both abducted from the bathroom of a large hotel and taken to a field. I was taken away. She was assured that I would not be hurt, and she was tied up in rags. The dream ended before anything else happened. I tried to reassure her that it was only a dream. She argued that it may have been a memory, or maybe a precognitive dream. I didn't have an answer for that. I was more concerned with convincing her it was only a dream, and reassuring myself. I wanted to believe that her mind had made this up, and had included me in it so that she would not feel so alone with her trauma.

As Beth related her discovery of her father's childhood memories, I started cataloguing the information. After all, I wasn't related and he was more like a character in a mystery novel to me; a mystery novel I wanted to figure out before the author told me her conclusions. Everything I'd read indicated that abductions were a relatively recent phenomenon, not appearing until the early sixties, yet I had information that indicated they may have begun much earlier. Her father's beach experience with his brother had to have taken place in 1928 or 1929. I knew then that I had information other researchers didn't have.

I was so fortunate to live with an abductee. All I had to do was get Beth and her father to remember more. Other researchers had to wait for their subjects to come to them at infrequent intervals to reveal small fragments of memory. I had my own 24-hour-a-day subject, and she had access to her father adding his own memories, and confirming some of hers. This confirmation of memories by another, semi-independent witness was a stroke of luck. Now I could really find out what was going on.

Unfortunately, my witnesses proved uncooperative. Beth's frequent denial of everything that I looked upon as fact frustrated my search for explanations. After all, what she told me was not dissimilar to what I had read about, but she didn't want to hear that. Her unwillingness to accept her involvement with alien beings, her bits and pieces of memories, frustrated me. Why couldn't she remember more? Was she just trying to test me, or worse yet, lying to me about what she remembered? This and her father's unwillingness to talk to anyone but Beth, left me groping in the dark again. All I could do was relate what I was hearing to Rob Swiatek and hope he would give me new insights on how to proceed. Beth wouldn't talk with him, but that didn't stop me. I needed someone to talk with to be able to sort out what I was hearing.

Beth has an active imagination, and I often wondered how much to believe of what she was telling me. It wasn't that I doubted that she believed what she told me, or that the basic incident had happened. I just didn't know how many of the details to believe. In the past she'd implied to me that I really shouldn't leave the horses out behind the barn after feed them (I do chores on Sunday) because they had ruined the outside of the barn by chewing off boards and destroying the doors. When I checked, I found a few teeth marks in the siding and a bent up roller on one of the doors. Not serious as far as I was concerned; normal wear and tear. A little paint and a hammer would fix it. Listening to Beth I had expected to find something destroyed. She and I have different concepts of significance. Rob and I didn't have that difference. He accepted what I said in an uncritical manner and tried to help me as best he could.

By this time I had collected so many books on the subject of UFOs, and a few on abductions, that my bedroom floor was a hazard to walk through, especially when I had to go to the bathroom in the middle of the night. So, I built a bookcase for the bedroom. I still couldn't think of keeping the books in one of the five bookcases we had downstairs. I still didn't want anyone to know I was interested in something this bizarre. They might think I was weird. That would be bad for business, and the depressed economy had done enough damage.

My father came to visit shortly after I finished the bookcase (filling three of four shelves with books), and my brother's bragging on it made him ask to see it. With trepidation I showed it to him, expecting the inevitable question about the nature of the books. I wasn't ready to tell him about any of this. Just because Beth had found the nerve to confront her father didn't mean I was that brave. I needed his love and I didn't want him to think I was crazy. I had enough problems with our relationship and couldn't risk that. I needn't have worried. He complemented me on the good job I'd done on the bookcase, looked at the contents, and made no mention. I was relieved, Beth was incredulous. Any normal person, and especially my father—with his insatiable curiosity about all things—would have made some comment. She found his lack of interest uncharacteristic. Her emotional side was overriding her logical self, or was mine?

My weekly calls to Rob Swiatek of the Fund were beneficial, but frustrating. I called him from work since I could do so without incurring long distance charges, but I also couldn't talk freely. I was always afraid of being overheard, so I couldn't say all that I wanted, and I had to keep the calls short to give the impression of working should someone drop by my cubicle. I also felt it was an imposition to call him at work. Just because I couldn't work didn't mean I had the luxury of taking him from his work.

I was excited when Rob proposed that Beth and I meet with other people knowledgeable about abductions at his house one evening. The more I learned about abductions, the fewer answers I had. I saw this as an opportunity to meet new people who might support me in my intense quest, and give Rob a break from my incessant questioning. As usual, Beth wasn't thrilled. Why did she have to act like such a ninny about this? Most of it was in the past, things were relatively quiet, and she'd survived intact. Oh well, I had my own agenda. I was going. She could come if she wanted to. She came.

23. Beth

Our inner circle was expanding steadily, more and more people insinuating themselves into what I had come to see as a private sort of hell. Although I was becoming more convinced with each passing day that Anna was involved in this mystery—perhaps as deeply as me—I had presumed she felt as secretive as I did. I should have known better than to presume anything with Anna! She had a mind of her own and there was no reason for her to isolate herself from other people just because of my reluctance.

Anna had an insatiable curiosity about the abduction phenomenon, which increased with every little revelation (or remembered mystery), until her personal library on the subject outgrew all available space. Books and magazines on UFOs, abductions, crop circles, cattle mutilations, New Age psychology, theory and other remotely related literature were stacked along her bedroom walls, on bookshelves (where space allowed), and on tables and chairs. Her inquiring mind brought her in contact with an ever increasing circle of acquaintances, including some famous writers, therapists, investigators and miscellaneous hangers-on. At one point she even referred to herself as a camp follower. I thought this label was rather harsh, but agreed that she seemed to be so wrapped up in the subject it was amazing she could function on any normal level!

For myself, I found it difficult to talk to anyone outside of Anna and my father. I had still not found the courage to confront my mother, finding that she would immediately remove herself from hearing range whenever my father and I brought up what she referred to as "that crazy stuff." I was not as uncomfortable speaking with Rob Swiatek as I had once been, and sometimes would call to update him on any new information, remembered experiences, flashbacks, unexplained scars which appeared overnight, missing time or family involvement. But as for other members of the Fund, I was still reticent to discuss these matters and even more reluctant to meet them.

So when Rob invited us to his apartment to meet some of the other Fund people, as well as a few interested parties, I was hesitant. When I learned that Richard Hall, a highly regarded author of books on this subject, would also be there, I felt somehow invaded! Did this mean we would be exposed to yet another demoralizing interview?

I did not wish to be cross-examined, no matter how famous the examiner!

Anna tried to reassure me that talking to others would be good therapy and would help us to sort out our feelings and put things into perspective. I knew it wasn't healthy for me to isolate myself, but just the thought of letting strangers meander around in my private torment left me feeling weak.

I finally gave in, after Anna threatened to go to the get-together without me. I can now admit that it was the right thing to do in drawing me out of my self-imposed seclusion, but getting through that first encounter was both embarrassing and frightening.

We arrived at Rob's apartment about 7:00 P.M. and were welcomed like old friends. Perhaps it wouldn't be so bad after all, I had thought. There were several people already there and we were promptly introduced to each in turn. Richard Hall, whom I'd recognized from a photo in a book, greeted us warmly. He seemed to be nonthreatening and I liked him instantly, already feeling more comfortable and forgetting about my inhibitions.

Bob Huff, a tall, broad-shouldered man in his mid- to late-thirties, introduced himself as an interested bystander, not a member of the Fund. He didn't seem to be an abductee and I wondered why he was there, already editing myself so that I wouldn't inadvertently say something revealing. If I was lucky, perhaps I could be an interested bystander as well. Seemed safer. Two women rounded out the small group: Sue, who claimed not to be an experiencer but interested in the phenomenon and a close friend of Rob's; and Carol, a social worker who was studying to be a hypnotherapist—another interested observer. Anna had mentioned speaking with Carol earlier in hopes she could fulfil our needs for a qualified therapist, but was dissatisfied with Carol's approach and hadn't felt comfortable with her. We both wondered, when given this type of forum, if the woman would prove more acceptable.

The first few minutes were spent simply socializing, getting comfortable with one another. I was starting to relax a little, enjoying being one of those "interested bystanders" I had heard so much about since our arrival. But the atmosphere soon changed when Anna's and my experiences were broached. Rob had briefly updated everyone there on our initial interview, and presumably on more recent events as well. As the questions about my sketchy conscious memories intensified, I found myself becoming defensive. I was not so much protecting my privacy (although that was certainly a concern, too), but rather I felt a strong mental warning not to reveal anything further. I didn't understand this admonition, but believed it was serious and I felt threatened by it.

When I told the group that I could not say any more because by doing so I would be in danger, they acted as if this response was normal and expected. I later learned that it was not unusual for an experiencer to perceive a real threat of bodily harm (from the aliens) when discussing certain aspects of an event, and that no matter how sincere the threat seemed, none of these reported threats had ever been followed up on.

Carol had broken into the conversations several times to ask her own questions and offer suggestions on how she felt victims of these abductions might resist. Her heart may have been in the right place, but her counsel was grossly inadequate, I thought. Her advice included such gems as, "Why don't you stay away from those places where you're being abducted?"; and, "Tell the aliens you don't want to go with them"; and, "Run away when you see them coming"; and, "Don't believe in them!" I was incredulous! These people were supposed to be *aware!* We knew next to nothing about this phenomenon compared to most of the people in that apartment, yet even we knew no one had yet to discover a successful formula for resistance. Gaping at the social worker, I mentally scratched her off our list of possible resources. She had a lot of developing to do before we could trust our psyches to her.

As it was getting rather late, Rob suggested we go for dinner at a restaurant nearby. The restaurant was located a few blocks away and was within easy walking distance from Rob's apartment. Our thoughts concentrated on filling our stomachs, we contented ourselves with trivial conversation on the way to the restaurant. I was relieved, having been concerned that the topic of abductions, aliens and UFOs would continue in public.

Seated with drinks served and dinners on order, the discussion again returned to abductions. I was mortified! There were people all around us, people who could surely hear every word. If there had been a hole under the table, I would have gladly crawled into it. How could they talk about this stuff right out in the open? Didn't they care that people would think we were all crazy?

I wanted to get away from them and from this insane subject, but I couldn't see any way to escape. I needed Anna to go with me but she was deeply involved in a conversation with Richard Hall and I couldn't get her attention. How was I ever going to get away? The situation was hopeless—worse than hopeless, it was humiliating!

Perhaps I was being oversensitive, but as I saw it, discussing such an off-the-wall subject in public did nothing for one's image—

or morale. I was already so upset by the implied threat not to reveal my experiences to these people that any more cracks in my already weakened protective wall would cause it to crumble into ruin. I could feel myself breaking up, the mortar disintegrating. Soon I would be nothing more than bits of rubble lying on the ground...and these people, these interested bystanders, would not be able to put me together again.

I don't know how I got through that dinner. I barely remember the drive back home or any conversation Anna and I might have had during the return trip. But I did survive it, and perhaps, over time, it helped me to accept the complexities of this phenomenon and those people determined to help its victims. One day, I told myself, those same people would be my friends, and maybe others I had yet to meet. If we worked together we might find solutions—or at least a better understanding. I would be satisfied with the latter.

24. Anna

My dependence on Rob had become almost obsessional. It didn't seem to be doing him any good either. On the days I was feeling matter-of-fact or inquisitive about the aliens, our conversations buoyed my spirits. The bad days, when I awoke exhausted and ravenous and phoned home to discover that Beth had felt the same way for no apparent reason, led to emotional conversations with Rob and the infrequent lunch. He would eat and I talked through my tears as I pushed the food around my plate, trying to make some sense of what was happening.

Neither of us was used to weepy women. I didn't want to continue to cause him this much distress, but I needed help. He handled it well, I must admit. Even when I offered to let him off the hook, I wouldn't talk to him about this as frequently, he refused to let me get away with it. He offered other options: Richard Hall and a new therapist who was interested in working with abductees. I didn't feel I could talk with Richard Hall. After all, he was famous. I'd read his new book *Uninvited Guests*, and his name appeared in almost every book I'd read on the phenomenon. I shy away from famous people. I feel uncomfortable around them. The therapist sounded like the better choice. She was also studying hypnosis, so maybe here was my chance to try it.

I telephoned the therapist, Carol, one evening when I was feel-ing particularly vulnerable. I knew I was imagining the parallels I saw between Beth's responses and my identical reactions. My mind was inventing the nausea, diarrhea, exhaustion and hunger on the same mornings as Beth. I wanted Carol to tell me I was stressed out. She didn't. The morning I yelled (in my mind) at a huge, brightly lit, square "helicopter" a few miles from the house to "Leave Beth alone!" and she fell out of bed at exactly the same time was hard to justify as stress! This was getting too weird for me. I wanted Carol to tell me I had an overactive imagination. She didn't.

Unfortunately, she knew less about abductions than I did, and her "California pop-psychology" platitudes did nothing to relieve my anxiety. Rob was right; she was new to the field, but I questioned her therapeutic knowledge as well. Maybe she was trainable, but I didn't feel like being her first guinea pig. Why weren't there more psychologists attuned to the phenomenon, that we could afford (ba-sically free), when we so desperately needed help?

Having a chance to meet Carol, and a few others, seemed like an excellent chance to re-evaluate my impressions of her. I looked for-ward to the March 29 meeting at Rob's. After the first half hour, I be-gan to relax. These were normal people, like Beth and me. Maybe not exactly like us, none admitted to being abducted or even suspecting it, but people I felt comfortable with; all middle-class, holding good jobs, and not espousing any strange philosophies. Just real people. I even got brave enough to ask Richard Hall to autograph his book for me. Rob had assured me he would be pleased to do so, but it took more nerve for me to ask him to do that than to go meet him.

Beth finally relaxed enough to discuss some of what was going on in her life. Her description of her latest encounter with the man in white with his outrageously large cowboy hat had us all in stitches. She could even see the absurdity of it. Rob and Sue suggested that they might go and talk with the lady that lived in the house where she called me from, to see if there really was a strange man in a white cowboy hat that wanders around Delaine helping stranded motor-ists. None of us really thought so, but there was so little that could be checked out where abductions were suspected, that any bit of data may provide another piece of the puzzle.

Beth went on to relate her father's beach experience in 1928 as if it were someone else's story. I knew this outing would be good for her! As she related the oft-heard tale, my mind went wandering to the family and genetic implications. I had met her granddaughter, Noel, a few weeks before and found her amazingly bright for a four year-old and very tall, with very light blonde hair even though both parents have brown hair. Had some genetic engineering been done? Was that even possible? Was she part alien? No, my mind was running away with me again. This was one of those disturbing thoughts that I dare not discuss even with Beth, no matter how close we had become. This was her granddaughter.

Beth's son, Paul, was also tall (6'5"). I'd never seen him, but if this kid kept growing she'd be over four feet tall before she was five years old. She'd just inherited a favorable combination of genes from her parents that are much revered in American society—intelligence and height. Being around her didn't disturb me as much as with most children, although I had little to do with her when she and her mother visited the farm. As usual, I managed to get through the introductions and then excused myself to go do chores, elsewhere. I just don't like to be around children. They're too noisy, too unpredictable, and they don't act like adults. Mentions of food brought me out of my reverie, and back to the discussions at hand.

During dinner, the conversation turned to women reabsorbing pregnancies. Beth leaned over and told me she had had a reabsorption and a miscarriage before her son Paul was born perfectly healthy. I said maybe they weren't miscarriages. She had not read that part in *Intruders* that deals with the missing baby syndrome. She then became quiet and later spoke to the group about it. Seems the second miscarriage was pretty messy, happened at home and there was blood all over everything. When she called the military doctor, they asked her to bring the fetus and placenta when she came in. She couldn't find a fetus.

What fascinating people, and they talk about all this stuff out in the open—even in restaurants and walking down the street. Feels good, but weird. I lost some of my isolationism that evening. All I could do was hope that Beth shared my feelings. I now had new people I could converse with, especially Richard and Bob Huff. Richard had all the knowledge accumulated in thirty-five years of studying UFOs. Bob had treated the subject with less reverence and seriousness than the others; I found it refreshing.

25. Beth

The spring of 1992 heralded a number of major changes in my world view. Beginning in April, I was besieged with unexplained vision problems that have yet to be understood or satisfactorily corrected. I had been wearing extended-wear contact lenses for about a year by then and had made a point of having my eyes examined every three to six months. I had not—before this time—had any problems with the lenses, but felt it was wise to have my eyes checked more often than I would otherwise have done while wearing only glasses.

Before one of these regularly scheduled exams, I noticed that I was seeing quite clearly even without my lenses and that my glasses were much too strong to wear comfortably. I moved my appointment up a month, worried that something was wrong with the prescription. The doctor made a thorough check, had me read some very fine print without my lenses or glasses (which I did without straining), but couldn't explain why my eyesight had improved so dramatically. I had been diagnosed with astigmatism in my left eye, yet when the doctor checked the eye again, there was no astigmatism present. We were both confused by this development. According to the doctor, astigmatism rarely went away once diagnosed, and that at my age it was perfectly normal to have at least mild astigmatism—often in both eyes.

Putting this enigma aside for the time being, the doctor proceeded to examine the underside of my lids for irritation, but found nothing unusual. He questioned a scar on the inside edge of my left eye, and I explained this was from an accidental burn when I was about ten years-old. He looked inside the lid and found what he described as "more recent development of scar tissue, within the past few months." I had no idea how that could have happened.

(Note: During a subsequent talk with my parents, I asked about the origin of this burn, relating my memory of having been burned when I swung my head around and stabbed myself in the eye on my Uncle Bill's lit cigarette when I was about ten. I recalled this having happened during one of our family's many camping trips. Surprisingly, my parents laughed! My father asked if I ever knew my Uncle Bill to go camping, and I had to admit it would have been very unusual, since my uncle's idea of roughing it was staying in a motel. He also reminded me that Uncle Bill had stopped smoking years before that. Still confused, I asked if they knew, then, where I had gotten the burn. They both admitted having no idea when or where I might have been burned in the eye.)

Over the next month, my eyesight alternated between nearly perfect (without corrective lenses) to what had been normal for me up until the previous month (requiring corrective lenses). I found it necessary to return to the eye doctor several times in an effort to correct the problem and stabilize the condition. The astigmatism has never returned and I have learned to live with the not-quite-right lenses.

In the meantime, Anna was having problems with her prescription. She was also wearing contacts and had noticed fluctuations in her vision as well. Making an appointment with another eye doctor who had no more success explaining the changes than mine had earlier, Anna found herself making weekly trips to have her prescription adjusted. But, as with mine, her eyes have yet to stabilize.

It seemed eye problems were the popular topic during that month. One of my riding students had cancelled a regularly scheduled weekly lesson because of an eye injury for which she had trouble finding a cause: Her husband had been working late and she was home alone when she decided to go on up to bed rather than waiting up for him. She was about to close the drapes over the patio door, which faced the back yard and woods, when she was suddenly blinded by a very bright light. Her husband arrived home some time later (she did not know when), and found her unconscious on the floor, the drapes still open. Having called the Rescue Squad, her husband concentrated on reviving her, but when she came to she was unable to see, complaining of her eyes burning unbearably. The Rescue Squad arrived and took her to the hospital where she was diagnosed with burned retinas. She explained to me that she was too upset to tell the doctors how it had happened. She was concerned that they would think she was crazy.

Several weeks later, I had called my student to find out how she was doing. I asked if her eyes were healing and if she recalled any more about that night and the bright light that had blinded her. She seemed confused, saying she didn't remember anything about a bright light, but was simply recovering from an eye infection! Eventually, she did take up her riding lessons again, but by then had no memory at all of the incident.

April was also the month of strange phone calls. During that first week, two calls were received which caused me to believe I was becoming paranoid. On the afternoon of April 6, I answered a call (using the farm name) but heard no reply. I kept saying, "Hello," assuming we had a bad connection and the calling party couldn't hear me. There followed a moment of absolute dead air, then a deep, brusque voice said loudly, *"Don't!"* I was so startled I nearly dropped the phone. The line went dead once again and I waited for the dial tone before hanging up, feeling unduly frightened by the one-word command. I couldn't understand what it meant and knew it shouldn't have bothered me so much, but I just couldn't shake the feeling that I was *expected* to understand.

Another odd call came on April 10. This time, again after a spell of dead air, the caller spoke in what I could only describe as gibberish. The voice was absolutely clear, each word enunciated with precision, as if I might not otherwise understand! I, of course, did not understand a word that was said. It didn't sound like any foreign language I had ever heard before; there were clicks and guttural sounds throughout. As this monologue progressed, I noticed a change in tone, as if the caller had become angry with me. Suddenly the voice fell silent and the line was immediately disconnected.

Two days before, I had received a call that was equally unexpected, but much more pleasant: My friend Dee, whom I hadn't seen or spoken with in almost fifteen years, had located me through my parents. She explained she had spent years trying to find me, that it had become an obsession and she was fortunate to have such an understanding husband considering the time and energy she had put into the search. I was surprised to hear from her and flattered that she would go to such lengths to find me, but I couldn't help but wonder why. Had something horrible happened that she felt I needed to know about? No, she replied, she just needed to find me! We spoke for some time, then promised to write and fill each other in on what had been happening in our lives.

Soon after, I did write to Dee, telling her briefly about the strange experiences, missing time episodes and memory flashbacks Anna and I had been going through. After she had received my letter, we again talked on the phone, and Dee amazed me by asking if these events had anything to do with alien abductions! When asked if she had been having similar problems, she said no, but that she had been reading about the phenomenon earlier and recognized the parallels.

Dee and her husband, a commercial airline pilot, came for a daylong visit on April 25. Dee and I spent the majority of our time reliving our teenage years together, looking at albums, talking about the mischief we made, the grief we caused our parents, and finally, the abduction phenomenon. Although we had not remembered it earlier, I asked Dee if she recalled the night her parents had called the police because we couldn't be found. She shook her head, saying her memory wasn't as good as it should be. I reminded her we had been about fourteen or fifteen when I had been spending the night at her house one weekend and we had decided to sneak out for ice cream at around 10:30. Dee's eyes suddenly widened.

"Yes," she said, "I remember now! We went out the back door from the rec room. Mom and Dad were upstairs and we didn't think they'd come check on us."

We were going to go down to the Dairy Queen and be back in less than thirty minutes, I recalled, but when we arrived back at Dee's house and were about to slip back into the rec room, we found the door locked. We found a downstairs window partly open and had raised it to crawl in when we overheard her mother's frantic voice on the phone. She was talking to the police and begging them to send someone out to look for us! At first, though we certainly knew we were in trouble, we could not understand why the police had been called. We had been gone less than a half-hour! To our dismay and utter confusion, we discovered later that it was almost 2:00 A.M.! Neither of us believed we could have been out that long and not have known it or been able to account for the time. Worse, neither of us remembered having bought or eaten any ice cream!

Dee then reminded me of our favorite collectibles: the furry little stuffed kittens. She had given each of hers different names, but mine had all been named Dee-Dee Kitten, though I had long since stopped demanding them as gifts. I couldn't explain to my friend why I had chosen that particular name years before I had met her. I didn't recall having ever before known anyone named Dee or Dee-Dee. Dee said she thought I had called her Dee-Dee when we met, but that I had shortened it to Dee after she'd objected to it, complaining that it sounded like a kid's name.

Bits and pieces of other troubling events were explored, but details were hazy and our memories of these incidents had faded over the years. In time, perhaps they can be retrieved.

(Note: Several months after this reunion, when I asked my granddaughter, Noel, what she would like for her birthday, she asked if I could buy her a Dee-Dee kitten! I had never mentioned these stuffed animals in her presence (or in the presence of her parents) and believed I had misunderstood her. I asked her to describe the kitten to me and she proceeded to tell me what I already knew— exactly what they looked like! Noel has never offered an explanation, and I have resisted the urge to question her further about it.)

With so much going on, so much clutter in our lives, I thought I wouldn't be able to handle anything more! Then, while speaking with my father over the phone later that month, I sensed he was angry about something, but couldn't persuade him to tell me what was bothering him. When I next went to visit my parents, I found my father still unusually distraught and finally convinced him to tell me what was wrong.

In angry tones, he described a visit by two men in Air Force uniforms who had dropped by unannounced on the morning of April 10. Explaining their visit as a routine security update on my son,

Paul, who had been promoted to Sergeant in the intelligence branch of the Air Force, my parents invited them in, expecting to be questioned about family background. Although neither my parents nor I had ever been questioned under this standard military procedure, I was not surprised by the Air Force's appearance. But my father was not upset by that, he explained further. He was annoyed because they spent the entire time grilling them about me!

According to my father's recollection of the conversation, one officer did most of the talking, insisting my parents tell them what they knew about my involvement with "organizations investigating anomalous phenomena." At first, my father went on to say, he wasn't sure what the man was asking, then when the question was rephrased, he had thought I had gone public with our family's recent discussions and was furious with me for doing so without his knowledge. I, of course, denied having done any such thing, and was finally able to convince him that the Air Force was, for some unknown reason, fishing for information.

But how could they know anything about this stuff? I'd not yet been able to bring myself to talk to my son or daughter-in-law about it, and other than relating my talks with my father to Anna, no one else knew any details of my childhood experiences. I had no explanation for the Air Force's interest in my experiences and assured my father that I had kept his confidence, as promised.

He seemed to accept this at last, and assured me he had revealed nothing to the Air Force either, telling them that if they had questions about me, I was an adult and no longer living under their roof so they could ask me directly and that he had no idea what they were talking about.

By the end of April, several of our boarders reported seeing more than the usual air traffic over the property. These sightings were mostly of helicopters hovering lower than would be considered both safe and legal, along with an occasional private plane circling the farm for long periods of time, also flying low. Once, a woman who helped us take care of the horses on Saturdays, reported seeing a small single-engine plane pass over the house several times while a man leaned out precariously and appeared to be filming! Others told us about hovering helicopters and low flying planes with box-like devices protruding from the fuselages. Although Anna, her sister and brother, and I observed some of these overflights, each time we tried to film one of the craft, they took off immediately. Particularly odd was the fact that none of these craft displayed any markings whatsoever; no numbers, letters or insignia that we could use to identify them in order to report their unauthorized and annoying flyovers to the FAA.

I was beginning to believe I would live through April after all, when I awoke on the 29th feeling nauseous. I could barely get out of bed, afraid I would throw up all over the carpet before I could make it to the bathroom. I did manage to reach the toilet before it was too late, and I did feel much better afterward, but while washing up I noticed blood on the front of my nightshirt. I almost didn't want to know where it came from, but couldn't resist looking anyway. I pulled up the shirt and saw a hole about a half-centimeter wide on the upper lip of my navel. It was sore to the touch, but had apparently stopped bleeding sometime during the night. At first I shrugged it off, not wanting to think it was connected to an abduction.

By mid-afternoon, I was obsessed with cleaning the house, feeling as if I were preparing for some important social event. When Anna questioned me about this behavior, I only told her I had to do it. It was important to me.

For the next several weeks I continued to feel sick in the mornings just after awakening, usually resorting to throwing up, then I spent every spare minute during the day cleaning house. I knew this was not normal behavior, but I also recognized what it symbolized: I was experiencing morning sickness and nest-building behavior! That would have been normal if I had been pregnant—and acceptable if I wanted to be—but neither applied!

In the fall of 1972 I'd had a complete hysterectomy because of lesions which were forming tumors on my ovaries and uterine wall. Although the cause of these lesions (scar tissue) was never fully understood, the surgery was certainly done, and there was no chance I could ever become pregnant again.

So what did these symptoms really represent? Anna and I discussed the problem and I decided to see a doctor. I made an appointment with a gynecologist for May 13 for a Pap smear and blood test, sure the pregnancy-like symptoms were the results of some ordinary illness.

The wait for a doctor's opinion seemed interminable. The morning sickness continued, as well as the nesting behavior, but other symptoms emerged as well; swelling abdomen, tender breasts and unusual weight gain added to my worries. I had been pregnant before and knew what these signs indicated, but my rational mind kept telling me it wasn't possible. Therefore, it had to be something else!

Then the day finally came and I happily imagined a diagnosis of stomach flu or bladder infection or gall stones or kidney infection or high cholesterol—*anything but pregnancy!*

I sat patiently on the examination table waiting for the doctor to appear, relieved I had not had to change into one of those backless hospital gowns. I was feeling a little nauseous, but assumed that was from nervousness. I had never felt comfortable in those stark white examination rooms being stared at by trays of lethal looking instruments and jars with suspicious contents. Finally the doctor came in; a woman! I was already more comfortable, although there would have been no reason to be uncomfortable if it had been a man. She smiled pleasantly, introduced herself, and looked me over carefully. Still smiling, she said cheerfully, "You look okay for being a good three months pregnant. May I ask your age?"

I told her, not even bothering to argue over her earlier remark. She seemed surprised by my answer, saying that although I didn't look my age, a pregnancy at this stage of my life could prove difficult, even dangerous, and had I considered the consequences? I told her I was certain I wasn't pregnant, had had a complete hysterectomy years ago, and hoped she could tell me what was really going on.

The doctor proceeded to have blood drawn and suggested I change into the dreaded hospital gown for a Pap smear and exam. There were problems almost immediately. When the doctor tried to use a speculum for the smear, she was unable to insert it because I was so swollen. I complained of the pain so she left to find a child's speculum, but this was nearly as bad. She was finally able to obtain a sample but seemed concerned about the inflammation and asked if I had been raped! I, of course, said I hadn't.

"The vaginal discomfort and inflammation is indicative of force," she explained. "Are you sure you haven't been sexually assaulted?" I again denied it, asking what else may have caused it. "Miscarriage, or perhaps a recent birth."

I shook my head adamantly. This wasn't working out as I had hoped! I needed another explanation for these symptoms, and obviously I wasn't going to get one. The doctor seemed to be as confused as I was.

I left the doctor's office in a foul temper, angry that she hadn't found a cause for my discomfort, angry that I wasn't able to confide in her about my problems, angry that I couldn't just make it all go away! My world was falling apart, my emotions all mixed up, my reality on shaky ground. Would things never go back to normal?

By the following morning I felt pretty good, for a change. I had no nausea and no urge to clean. It was wonderful! I started to get dressed, delighted to find I could get into my jeans without holding my breath. I examined my breasts and noticed they were no longer so tender, but the cut on my navel had apparently bled again during the night. Oh well, I could live with that.

Suddenly my mood shifted and I felt something was wrong. I sat on the edge of the bed, wondering how I could be so content one minute then so depressed the next! I thought back over the previous night, remembering that I'd woken up about 3:00 A.M., but I couldn't at first recall why. I took a deep breath and concentrated. Then I remembered! I awoke feeling as if I were being watched. I sat up in bed and looked toward the door, seeing something on the other side of my dresser. Thinking it was probably a shadow, I waited for my eyes to adjust. Then the shadow moved and I recognized one of the little greys (the ones I later referred to as the escort service) peering around the corner of the dresser. I was terrified, not remembering ever having seen anything like that before! Strangely, I was suddenly quite calm and relaxed, thought to myself that it was only them, and went directly to sleep.

Telling Anna about this memory later that same morning, she told me about having awakened around the same time to find two small greys standing next to her bed looking down at her. She said she hadn't felt afraid, hadn't even considered that it might be unusual to find them in her bedroom in the middle of the night! When she reached to turn on the light, she fell back to sleep again instead, not waking up until after dawn.

We made a point of entering these latest events in our journals, along with the results of my visit to the gynocologist. The tests had come back "normal," all indicators within acceptable ranges, yet I was not necessarily relieved. If a doctor couldn't find out what was wrong—if it wasn't a medical problem—then how were we to explain these things?

May crept along and we followed with trepidation. What else would disrupt our lives this spring? It had become increasingly difficult not to react to every bump in the night, every coincidence, every dream, and not associate these things with the phenomenon. We struggled to keep up with our journals using the word processing program on the computer, but were even paranoid about that benign activity! We had noticed parts of my journal files were being destroyed (or lost) and blamed it on the computer. The hard drive was probably going, but we couldn't afford another one right away.

One day, with Anna's brother, Rick, as witness, several paragraphs of my journal entry disappeared, one line at a time! The whole file was not removed, only selected portions, and I couldn't believe that a bad drive would not have destroyed the entire file, rather than segments. The only information left on the file was remarks not related to UFOs or abductions. These mishaps continued until the modem was disconnected. Spies? Paranoia? Who knew? We certainly didn't!

As a welcome change of pace, my son and his family made arrangements to spend Saturday, May 23, visiting me at the farm. Noel was already showing an interest in horses and had even ridden my mare on two previous visits, so it looked to be a pleasant day for all of us. After spending some time with the horses, we returned to the house to relax and talk. Noel seemed bored with our adult conversation, so I offered her some colored markers and poster board to keep her occupied. Paul and Sandy and I talked among ourselves while their little five year-old daughter sat on the living room rug, busily creating a masterpiece.

We were interrupted a short time later by Noel, who wanted to show us her drawing. When she held up the poster board to show me, my heart jumped into my throat! She had drawn a large triangle in the center with a light in each corner, which she said was a "flying machine." She described it as having windows around the edge where the people inside could look out. Drawn inside the triangle was what appeared to be a kite. She explained that she'd drawn the kite to show that the machine flew as quiet as a kite. To the right of the triangle was a figure with an oversized head, large black eyes and long thin limbs. The fingers were short and stubby (three on each hand) and the feet had squared-off toes. She identified this figure as "Nu" (spelling is Noel's), and said he came into her room at night, after her parents were asleep, and took her to strange places. Sometimes they rode in the flying machine, she added nonchalantly.

Below Nu was a box with another smaller figure inside. When asked who was inside the box, Noel replied it was her, that she sometimes wished she could have a box to hide in, so that when she didn't want to go with Nu, she could lock the box and he couldn't take her with him. On the left lower corner of the poster board she had drawn a tube-like outline with herself again inside. This, she explained, was a tunnel that led to Nu's home; it had green stripes on the ceiling and the whole tunnel was lit up with a reddish glow, but she couldn't remember seeing any red lights on the walls or ceiling. (See Figure 2.)

I didn't want to believe what I was hearing, didn't want to associate it with alien abductions, but I couldn't find any better way to describe it than Noel herself had done. My granddaughter had a wonderful, active imagination; but that alone would not have justified these drawings—or her descriptions that so closely resembled my own abbreviated memories of these beings!

Figure 2. The picture drawn by Beth's granddaughter at age 4.
(Photo courtesy of C.D.B. Bryan)

Her parents had also seen the drawing and listened to their daughter's descriptions. I looked to them for an explanation, hoping they would tell me it came from some movie she'd seen or book she'd read. Paul only said she'd been drawing "Nu" for a long time and often taped the drawings to the wall over her bed. Paul admitted tearing most of them down because, as he put it, "They scare her and give her nightmares." Sandy looked askance at me, suggesting it was the father who was upset by these drawings rather than the child. Although Paul denied that the drawings unnerved him, I instinctively felt otherwise. He was afraid! But why?

I remembered a number of incidents when Paul was very young, about four or five, when he would awake in the middle of the night screaming that "the cats were looking at him through the window." I had always believed these to be normal childhood fears and nightmares, but suddenly I remembered telling my young son not to be afraid of the cats, that I would go with him when the cats took him away! Why hadn't I remembered that before? It was clear at that moment what I'd actually said and done in response to Paul's nightmares.

I also remembered *not* responding to Paul's cries for help. There were times when he would scream for me to send the cats away and I was unable to get out of bed and go to him. I recall being "told" that I was not needed and to go back to sleep—and I did. In the morning we never mentioned the night's terrors.

Not willing to discuss such things in Noel's presence, I changed the subject (to the parents' mutual relief), promising myself that I'd get up the nerve to tell Paul and Sandy what I'd gone through. I did not want to repeat my father's mistake by hiding my experiences on the supposition that Paul had no memories, so why upset his world-view? Now that Noel might also be involved in this mess, I felt responsible for telling them everything I knew. I realized that Paul's realistic nature would probably result in total denial, but I was an expert in denial and recognized the signs! I didn't, however, want him to think his mother was crazy. But if it helped Noel, I was willing to take that chance.

Now I was afraid for Noel, for Paul, and for myself and Anna. This was becoming so complex! How would we ever unravel it?

chapter 6

M.I.T.–Close Encounters

26. Anna

Richard Hall's invitation for Beth and I to attend the Abduction Study Conference held at M.I.T. in June of 1992 was a turning point for me. It was a closed conference designed for invited investigators, therapists and abductees as an opportunity to explore the abduction phenomena and develop support systems for investigation and knowledge.

I knew that Richard considered me to be an unknowing abductee, but I felt that the only reason I was invited was so that Beth would attend. We both knew that she would never attend without me. It took two months of not-so-gentle cajoling on my part to even get her to agree to go. I was shocked when she eventually agreed to be on an abductee panel. I also agreed to be on a panel, even though I knew that I was a fake. I wanted to go so badly to gain the latest information, I agreed to relate my experiences with electromagnetic effects, shared dreams with Beth, my bedroom visitor and the feelings I was having about exploring my possible involvement with the aliens. That was quite a list, but I didn't think any of it was that important. But, it was enough for Richard, so we prepared to attend with trepidation. It was one thing to talk among ourselves, but to actually tell strangers was frightening.

The conference was all that I hoped it would be, and more. I met, or at least saw, many of the people whose work I had been avidly reading for six months. But more importantly, I had a chance to talk with other abductees about their experiences and how they managed to live with the knowledge. Beth and I were probably the most recent abductees, in terms of finding out about our possible involvement. Many others had been dealing with that knowledge for several years (one for more than twenty years), and seemed to be coping well. All of the abductees had undergone some hypnotic regression, something that Beth and I hadn't been ready or able to do.

For the abductee panel I had prepared written materials for publication in the proceedings that explained my feelings and what had been happening to me up until that time. I have reprinted that material here.

Psychological Effects of Shared Experiences

Beth and I first met in September of 1987 when she applied for a job as farm manager at my horse breeding business. I had several applicants for the job, but when Beth arrived I felt that there was something special about her and hired her almost immediately. It seemed as if we shared a common bond. I assumed at that time that it was a love of horses and an empathy with these magnificent creatures that I entrusted to her care. I now think that there are other shared feelings and experiences that may have brought us together.

As the years passed, Beth and I became close friends, but we are much more than that. We have a greater understanding of each other than anything I have ever felt with anyone before. We have very different personalities; hers is outgoing and people-oriented, while mine is much more reserved, analytical and suspicious. We make a great team on the farm. Between the two of us, we can handle any situation that arises. Yet we are also very much alike in other ways, so much so that some people have a hard time telling us apart, even though we look quite different. We have developed a synergy that I have never experienced before. Many times we confuse other people during conversations because we don't finish sentences or we switch to different topics without any intervening words, without any loss of our train of thought. We finish each other's sentences, and often will say the same things at the same time to each other or to others in conversations.

Our first shared experience with UFOs came on an evening in September of 1989 when we saw bright lights in the sky over the barns. It wasn't until January of 1992, when we became deeply involved with exploring Beth's recent abductions, that we realized we had two very different memories of that experience. We both remember seeing the two sets of brilliant lights from the back deck, but we do not agree on how many lights there were. Beth remembers watching them until one light from one of the groups sped off over the house, and later, they all receded swiftly from sight. Beth remembers my being with her all the time on the deck and then coming inside, talking about it, and being very excited about what we had seen. I remember seeing the original lights and then, uncharacteristically, going back inside and watching television until it was over. I have a clear memory of seeing the single light depart from the cluster, but no memory of how they disappeared. If I had gone back into the house, I couldn't have seen the one depart. I carry both of these memories in conflict and have not sorted out why my mind is doing this to me.

Since Beth's recent abduction experiences began, we have also had many unexplained shared experiences. One involved my bedroom visitation on March 13. If it was a dream, it was very realistic. I awakened during the night and turned over to see a large man looking at me from the side of my bed. He had to be at least six feet tall, broad shouldered, with a very broad top to his head. He was in shadows, illuminated from the nightlight in the hallway and maybe from starlight coming through the windows. I saw an outline only, but he was very solid looking. I remember asking out loud, "Who are you?" I don't remember any answer.

I reached for the light at the head of my bed and when I turned it on, there was no one there. I didn't feel fear, just curiosity; I'd finally get to ask all those questions I had about them. Strange that I didn't think of him as a burglar, or why I wasn't afraid that he had come to harm me. When I turned on the light my clock read 4:45 A.M. I turned out the light and immediately fell asleep. That morning I asked Beth what had happened the night before. She said it was busy and that the dog had left my bed several times during the night. The electricity in her room went out at 4:15 A.M. though there was no power failure anywhere else in the house. There may be a half hour unaccounted for from the time the power failed until I turned on my light.

While returning home on April 29 from a meeting in town about 10 P.M., we had an interesting experience with the lights on the freeway. We were driving the same car that had a complete electrical malfunction a few weeks before when Beth had an hour of missing time. As we passed a light pole on the interstate, it went out with a flash. About five miles down the road a whole bank of lights, about an eighth of a mile stretch, all went out. When we passed the next light that was lit, that whole bank of lights went off, too. About ten miles down the road we passed an interstate weigh station for trucks and the lights came on as we passed.

Another incident happened in the early morning hours of May 17 when we shared very similar dreams. Beth's dream was a dream within a dream. She dreamt that she woke up from a dream and saw a small gray figure standing in her doorway peeking around the dresser at her. She thought nothing of it and went back to sleep. She woke up again within the dream and saw two gray figures at the end of her bed. She went back to sleep immediately after seeing them.

My dream involved seeing four little gray guys just standing there. My only recollection was, "Oh, it's them." It was 4:30 A.M. when I awoke from that dream. We have no other memories of the dreams. The power went out in Beth's room that morning at 4:15 A.M.

On May 19, Beth and I were talking about the unpublished symbols that researchers use to validate abduction experiences. We talked about the doodles that we both had been drawing since childhood. We had talked about this subject before, but for some reason hadn't drawn our favorite doodle for each other. When Beth showed me her doodle, I was astonished. I then drew mine for her to see. The doodles are the exact same thing, except one is upside-down.

I've had a difficult time trying to understand what's been happening to Beth, and myself by association. I have no remembered abduction experiences, yet there are questionable times and events in my background that I am no longer able to satisfactorily explain. The recent events have caused me to question my sanity, my old patterns of explaining events and my sense of what is real and what isn't.

Since January I have been obsessed with trying to find answers, and all I find are more questions. Without the help of Robert Swiatek and Richard Hall, and their cautions not to become obsessed, I would be in much worse shape psychologically than I am. Trying to figure out what is going on is the last thing I think about when I go to sleep, and the first thing I think about in the morning when I awake. Did something happen last night? If it did, do I have any memory of it? Was it really a dream? Is today going to be normal, or will there be another bizarre event or revelation? I just don't know now. I used to have trust in my own abilities, be able to make judgments, but I can't anymore. Most days it's really hard to maintain a sense of normalcy.

During the day it is very hard for me to concentrate and get work done. What we have been experiencing seems to surface all the time. At work and at home, deadlines for projects all slip away. I'm lucky that in my job I am a policy analyst and have different projects to work on; my time is my own. I have been harassed a little about not meeting any deadlines lately, and have found believable excuses for my boss, but that won't last much longer. I feel guilty about not being able to concentrate on the projects, but that doesn't get them done any faster. I finished one this week (a month late) that my boss thought was good, but it's a poor piece of work. My other main project should have been finished two months ago. At least now I have quit taking books on UFOs and related subjects to work with me. Now I use my break and subway time to read work-related literature.

Until very recently I hadn't been able to concentrate on my horse business because of these incidents. The obsession is starting to fade, but only slightly. I still have many obligations to fulfill; I have a magazine to edit that is due in July, however I don't have any articles prepared. I just can't seem to get anything organized anymore. But this is a deadline I can't let slip. I have not been able to ride and train the young horses; I can't seem to get up the energy and enthusiasm for it. It's hurting business since I can't sell unrideable horses.

Beth and I have each other to talk to, and usually one of us will play the devil's advocate, but sometimes we seem to feed on each other's fears rather than help each other cope. Rob and Richard are great listeners and have helped me to keep things more in perspective, but sometimes that's not enough. Last February, when I was really feeling depressed and frustrated by all of the questions, I asked Rob for help from a mental health professional. I needed someone else to talk to that might be able to help me get my life back in order. I didn't like the uncontrollable mood swings I was going through. One day I would be ecstatic and cheerful, the next it was all I could do to keep from crying all the time. Too many other days I just felt emotionally and physically drained.

Unfortunately, Rob and Richard had no one in the local area that they had worked with in the past. I was put in touch with someone locally who might be able to help me. I spent an hour talking to her one night on the phone, but that didn't make me feel much better. Unfortunately, she had less knowledge about the phenomenon than I did, and kept talking in platitudes—"whatever you think is real, is real for you"; "you're very strong and brave"; "it's okay to feel this way." I didn't feel I got any help, and will not talk with her again. More than ever I feel the need to integrate these experiences to resume a coherent, productive life.

I feel that I have put an unfair burden on Rob and Richard. I'm not sure they understood the emotional investment they would have to make in us when these things started. Neither did I. When I first made contact with Rob, through the Fund for UFO Research, I thought that they would be able to give us a rational explanation and it would all go away. Obviously no one can make it go away.

But I have to rely on them for a lot of my logical and emotional support right now. It's such a big secret; I can't talk to friends about this. I don't want them to think I am crazy. That is one of the hardest things to deal with; it's such a big part of my life, yet I can't talk about it. I need to talk to a good therapist about this, but one who can accept the possibility that all this may be real.

I want to sum up this commentary by expressing the urgency I feel. Instead the first thing that now comes to my mind is, "Please don't think I'm crazy."

I felt what was covered here was sufficient for other people to believe that I was an abductee, and give people unfamiliar with some of the associated phenomena useful knowledge. It helped convince me that something was going on when the night before our panel discussion, as we drove back to the hotel, several banks of lights on the freeway went out as we passed.

27. Beth

An Abduction Study Conference? Who were they kidding?

When Dick Hall called to tell us about an abduction conference scheduled for June at M.I.T., I laughed aloud. What was there to confer about? No one knew anything; no one could explain anything; nobody had any realistic suggestions on how to stop these intrusions.

Who would be going to this conference? I asked Anna afterward.

Investigators, researchers, scientists, therapists, medical doctors, and, of course, experiencers (both abductees and contactees), she replied. We had also been invited.

What for?

So we could find out more about what was happening, meet others who were going through the same traumas, and maybe even make contact with therapists who could help us deal with it. This, apparently, seemed like a golden opportunity for Anna!

I didn't feel the same way. Why should we go all the way to Boston to expose ourselves to ridicule, when we could be ridiculed right here at home? Besides, I had no interest in being confused even more by the Doomsday theorizing of self-proclaimed contactees! I had limited knowledge of the claims of contactees, it was true, but from what I did understand, it appeared they believed themselves to have been chosen by these aliens as emissaries. These chosen ones were to spread the word, so to speak, of an imminent world-wide annihilation—an ending that would eventually assure the survival of our human race. How this would be accomplished was vague, but believers proclaimed that the aliens were only trying to help us save ourselves, that they had the most sincere and selfless motives, and that their interference in our society and evolutionary process had been going on since before recorded history. The role of these contactees was to relay this alien message to the rest of the world. "Space Brothers" was a term repeatedly connected with this camp. An interesting hypothesis, but a little too bizarre for my taste!

Anna seemed taken aback by my disinterest in attending the conference, saying that if I wouldn't go, *she couldn't*. "And why can't you?" I asked her.

"Because *you're* the abductee," she said pointedly, as if I had knowingly committed an embarrassing social faux-pas. "The only reason I was invited was so that you would go."

I didn't believe this, still feeling strongly that Anna was just as involved, but I promised to think about it anyway.

In May the invitation from M.I.T. arrived, along with a tentative schedule for the four-day conference. Included in this outline was a planned Abductee Panel, hinting that organizers of the conference would like to have "experiencers" volunteer to sit on this panel. *Ho boy!* I thought. No abductee is going to voluntarily expose him or herself to a few hundred strangers! They'd never get anyone to agree to that, I was sure—except perhaps for the contactees; they had already exposed themselves.

Soon after the arrival of the invitation, one of the conference's organizers called to ask if Anna and I would be willing to join the Abductee Panel. Another opportunity for a guffaw! I was assured that our anonymity would be protected, our true names not used without our permission, and that what we decided to reveal during our few minutes in the limelight was entirely up to us. Guffaw, guffaw!

I tried to explain to the organizer that Anna and I were still very new to this business and I, for one, didn't feel there was anything I could contribute to the panel or the further understanding of the observers. We hadn't even decided if we were going to attend!

Here I must add that only *I* had not yet decided. Anna was determined and I was almost convinced that it couldn't do any harm to go—but that was as an observer, not a participant! It didn't seem fair to deprive Anna of the experience, since she was obviously very excited over the prospect, and I was pretty sure she wouldn't go without me. So I agreed to go. How bad could it be?

And the Abductee Panel? Well, since we were guaranteed anonymity, and it was highly unlikely anyone other than Dick Hall would know our true identities, perhaps it would be an ideal forum for educating the therapists and encouraging their participation. Therapists knowledgeable about the phenomenon were in short supply, we all agreed. This might be the best way to gain access to them.

The panel participants were asked to submit typewritten material in advance, describing the planned topics of discussion. These would later be included in the published proceedings scheduled for release to the conference attenders later in the year. Although my original submission was sent as requested, the computer file copy I had planned to reprint here was mysteriously deleted from the data base during the "computer tampering days" (as described in a later chapter).

Briefly, I wrote of my family's connection with the phenomenon; about my father's, uncle's, son's and granddaughter's similar experiences. I wrote of Noel's description of her poster board drawing and the puzzle of my entire family's apparent involvement. Writing all this down was not as difficult—or traumatic—as I had expected it to be, but then I did not take into account that I would have to relate these things aloud in front of a room full of total strangers—believers and skeptics alike. I would be in for a real surprise, so it was probably best that early on it seemed to be a simple, harmless exercise.

We arrived in Boston a day early. I wanted to drive around and check out the old neighborhood. I had lived in the Boston suburbs from 1972 until 1981 but hadn't been back since. It felt good to see how little had changed, but even so I had trouble getting around with any confidence. After a few hours sight-seeing, we drove farther west of the city and found a quiet, inexpensive motel not too far out for the daily commute to M.I.T.

The next morning we made an appearance at M.I.T., registered and picked up our name tags. (We had agreed to use only first names for identification purposes.) I didn't see anyone I knew, but Anna recognized a few names from her many books on the phenomenon. Feeling stiff and out of place, I found myself searching for a familiar face! I knew Richard Hall was supposed to be there, but I didn't see him initially. I did notice that Anna was tense as well and wondered if my uneasiness had rubbed off on her. I was truly hoping this experience would provide the evidence I so wanted to find: That I was not an abductee; that I was probably crazy; and that Anna was suffering from sympathy pains.

I was continually wafting back and forth on the subject of insanity! Some days I couldn't discount the memories, my family's connection with those memories, and my feelings about Anna having been my friend since childhood; then, usually after a lull in the intrusions, I would again wish for insanity, deny anything unexplained had ever happened either to me or to my family, and embrace the belief that the connection was nothing more than hereditary mental illness. I theorized how easy it would be to commit myself for treatment. All I need do was convince a psychiatrist of my insanity and then I could relax and concentrate on getting better. It would be so nice not to have to worry about this stuff any longer! People would take care of me, keep me out of trouble, give me pills to dull the psychic pain, and erase the little gray shits from my mind! That was the only way to get rid of them once and for all!

But then something would happen: An unexplained cut or puncture, a fragmented memory of having been floated *through* my bedroom window, a shared "dream" with Anna. Whoosh! I was back in that other reality yet again.

Being at the conference was going to make this uncertainty even harder to control.

I noticed how many experiencers were registering during that first morning and felt as if I'd dropped into a scene from *Star Trek*. No...that's not quite right. These experiencers were just like us; normal average people who looked as uncomfortable as I did. They seemed just as insecure, just as nervous. This wasn't a scene from some futuristic movie, this *was* the future.

The Abductee Panel convened on the second day, and Anna and I joined them at the long table set up in front of the lecture hall. I was as jittery as I could ever remember being, thinking I had made a fatal mistake by coming to this conference. As each of us were introduced to the audience, I kept my eyes diverted, not wanting to see those wolfish faces glaring down on us. I had learned there were some invited reporters in the audience (though we had been assured no cameras would be permitted), and envisioned tiny spy cameras trained on the panel, clicking away and recording our faces (and identities) for posterity.

As each of the experiencers in turn told their stories, contactees included, I realized we each had developed our own ways of coping with the intrusions into our lives. I had relied on the insanity fallback; others dealt with the trauma by seeing the aliens as "space brothers"; some concluded they were experiencing a religious phenomenon with the attending angels and devils; still others saw the intrusion as an organized and self-serving exploitation by beings who had learned to manipulate space-time. There were, in fact, more theories than one could shake a laser beam at!

Since others were to speak before me, I took the opportunity to glance over my notes. It struck me then that I would have to say these things out loud! I didn't think I could do that with any semblance of self-control. I fingered Noel's poster which stood on its rolled edge on the floor next to me, realizing I had planned to show the drawing while describing what the figures represented to her. How could I do that?

Suddenly it was my turn. The room seemed very quiet; I could hear a man cough in one of the upper rows; a woman on the panel cleared her throat. I dared a glance up and came face-to-face with the wolves! They were all looking at me! What were they expecting from me? A performance? I hoped not. I would simply read from my notes and pretend there was no one there to hear. I could do this.

I began by reading, peering up from my notes occasionally to make sure no one was laughing. It was going fairly well, I thought confidently. Then I reached for Noel's drawing and opened it up for the audience to see, intending to point out specifics as I read Noel's descriptions. I don't now recall what particular word or phrase caused me to break down in front of so many strangers, but break down I did. Unable to maintain my composure, I buried my head in my hands and choked back tears of pure agony. I had never felt so inadequate! My granddaughter was telling me, by her explicit drawings, that she was being taken against her will by beings that defied description—or belief. And I couldn't help her! I couldn't save her from them, couldn't take the memories away or stop it from happening. I was useless to her, as helpless as I had been with my son. Was this how my father had felt?

God! What could I do? What could any of us do?

28. Anna

Recent incidents I dared not reveal at the conference were the pregnancies that Beth and I both seemed to have had. For several weeks in April and May Beth had been acting strangely pregnant: morning sickness, nesting behavior (cleaning, straightening up the house, washing windows, etc.), feeling bloated, and gaining weight only in the belly. This started a few weeks after she woke up one morning feeling as if she had been sexually abused. She also had strange dreams about a black woman trying to get her to take some pills so that she wouldn't get pregnant. That was crazy—Beth had a complete hysterectomy when she was twenty six! But when two home pregnancy tests showed that her body was manufacturing the hormones indicative of pregnancy, we got scared.

After several weeks we were able to talk over the telephone with a Canadian doctor who was familiar with the phenomenon. He was reassuring if not too helpful, but he gave us some strategies for approaching doctors and getting things checked out. By this time her stomach was so large that friends were beginning to comment on it, her nesting behaviors were getting her down, and she was very scared. Beth's doctor's appointment on May 14 was inconclusive—she couldn't get up the nerve to ask the doctor to run a pregnancy test. The doctor did find such extensive vaginal scar tissue and abrasions that she asked Beth if she had been raped. The doctor also jokingly remarked, when she saw her stomach, that Beth was in pretty good shape for being three months pregnant! The doctor found a slight vaginal infection, blood in the urine, and a 102 degree temperature—all of which Beth was showing no signs of having. So we didn't gain much besides confusing a doctor, a pattern that has continued with many doctors over the years.

After the May 18 shared dream that I mentioned in my remarks at the conference, the part I left out was that when all of Beth's symptoms of pregnancy disappeared overnight, I got them the next morning! I started having morning sickness, cleaning house (nothing I hate worse, and I do it only if I am embarrassed into it), and all my pants became very tight around the middle. The worst thing about it was I developed breasts! Very sore, large breasts! Me, who hasn't worn a bra in over twenty years. Never needed one, couldn't pass the pencil test.

By May 26 I couldn't ride a horse, or run down to the barn without severe breast pain—too much bouncing! I borrowed Beth's new bra that she had just bought for the same reason. She didn't need it any more. The home pregnancy test kit showed that I wasn't pregnant, but I sure felt that way. I have since learned, by watching the Oprah Winfrey Show, of all things, that when women are implanted with an embryo, pregnancy tests will show them to be not pregnant.

I started my period that afternoon, but my menstrual flow did not look normal when it started—too bright a red color and very thin, like blood rather than menstrual fluid. The next day it looked more normal, but the accompanying cramps left me feeling pretty rotten. The cramping was so severe the next day that I didn't go to work, something I hadn't done in years. The bleeding stopped the next day, but I still had cramps, felt bloated, and continued to spot blood for the next week.

I finally went to a doctor on June 6. He found a small mass in the right side of my uterus, and did urinalysis and blood testing for pregnancy—they all came back negative. The doctor thought I might have fibroid tumors in the uterus, so he scheduled a sonogram for two days later.

The night before the sonogram, I picked up on a telephone call where there seemed to be no one on the other end of the line (not uncommon by then). I listened for fifteen to twenty seconds, then hung up the phone. Within half an hour I was falling asleep in my chair in front of the television even though I had felt alert before the call. Though it was only 8:30 P.M. I went to bed and slept soundly until 7:30 the next morning. I went for the sonogram at 9:30 A.M. They found everything normal and there was absolutely nothing there, not even fibroids. The cramps and spotting disappeared after a few days and I started a normal period eighteen days later. I was just relieved that the pregnancy was gone, even if it had been all in my head.

One of the most important talks I had with another abductee at the conference at M.I.T. was a woman who had experiences since childhood. I had talked with Gloria on the telephone, but the conference gave me a chance to more freely express some of my fears. I have a phobia that only people who know me well know about—I hate and fear small children. It's not something that is socially acceptable to talk about, so I hadn't, but if one were around me any length of time, it would be hard to disguise. If a small child comes into the room where I am, I leave as quickly as possible. When my sister and her young son came to live with the family for a few months in the middle 1980s, I found myself depressed, and seriously considered building myself a small cabin on another part of the farm so I could live by myself. Many unpleasant fights ensued when my parents and other sister wanted them to stay with us, and I kept saying I would never permit it—I would leave the farm before I would consent to that. Pretty strong reactions. I put it down to depression and stress at work. Gloria had the same phobia towards children!

Gloria said that her phobia came from having to deal with all of the hybrid children on a ship while she was still a teenager. She was forced to hold the babies and nurture them for hours at a time. She was being unknowingly abducted so often during those years, was so tired, that she "slept"for twelve or more hours a night. After this intensive exposure to the hybrid children, she could no longer deal with real human children. Her sleeping for twelve to fourteen hours a night triggered my own memories of sleeping like that for most of my high school years. I'd sleep for nine or ten hours on school nights and then have to sleep till noon on Saturdays and Sundays to catch up. I always assumed I needed that much sleep because I was so physically active (swim team) and always anemic. Maybe not. I found it unnerving that we had the same phobia, and that Gloria's was directly related to her abduction experiences. I didn't like the implications, but still wasn't convinced I was an abductee.

Other abductees' experiences did not disturb me as much as Gloria's. Many described experiences that I couldn't relate to, let alone convince myself that I too was experiencing the same thing. One thing I did find fascinating was the variety of coping mechanisms that each victim chose to maintain sanity. Some had accepted the aliens as friends and protectors; they felt the aliens were helping humankind to deal with environmental problems and avert nuclear holocaust. Some felt that they had given the aliens permission to abduct them, and were doing this willingly for a variety of reasons.

I relate both of these coping mechanisms to the Stockholm syndrome. This way of maintaining a shred of sanity arose from studies of returned prisoners of war. They believed that their captors were friends who had their best interests at heart in the POW camps, and the captors became the good guys in the war. The prisoners attached their allegiance, and sometimes even their love, to the people who treated them so atrociously. I saw the same thing happen to my college roommate who had been abducted, raped and held prisoner for three days by a bank robber. She really didn't want to testify against him at the trial. I can understand the emotions that accompany this coping mechanism, but it wasn't something I needed to accept for myself.

I admit I felt closer to some of the other abductees who hated the aliens, hated what the aliens did to them, hated them for their unwanted intrusion into their lives, hated them for taking control, and hated them for destroying family relationships. Some abductees wanted to do bodily harm to the aliens. Maybe this is why the aliens usually keep most abductees in a near catatonic state, except children, whenever they take them. Then there were the abductees who believed that they had been aliens in a previous life and therefore felt a closer bond with the aliens.

My mind had been opened to many ideas and possibilities that I had always considered untrue before, but this was way too much for me to swallow. I believe that there are as many ways of coping with these intrusions as there are abductees. I'm for adopting whatever mechanisms allow one to function quasi-normally in this reality. Paraphrasing Budd Hopkins, "99.9% of our lives are lived here."

At breakfast one morning at M.I.T. I had my first hypnosis session—unknowingly! I was talking with a well-known California investigator who asked me if the aliens that were abducting me were the same group that he dealt with. He told me that there were several groups of aliens involved in abductions, some more considerate than others. The ones he dealt with were kind. I didn't really believe him, so I asked how he knew that to be true. He replied that it would be easy to find out. He asked me to close my eyes, look up at the ceiling and use finger responses to answer some questions. In that way my subconscious would answer the questions. He chose a different finger for me to raise to respond to his questions indicating yes, no, don't know, and don't want to answer. This seemed harmless enough, so I agreed to try it. It seemed to work. As he asked questions, my fingers would raise of their own volition in response. I noticed that the breakfast table noises seemed to diminish after I closed my eyes and listened to his softly spoken, rhythmic voice. I really had to concentrate on hearing the questions. I don't remember all he asked me, especially those that I signalled I didn't want to answer. The ones I do remember are:

Were you abducted as a child? (yes)

Was it frightening? (yes)

Did you get over the fear? (yes)

Do your aliens know me (his name)? (no)

Were these aliens the same ones that Beth was involved with? (don't know).

He asked the same questions using the other hand to respond. The results were the same. I found it an interesting experience, but not one that I felt confirmed any real abductions. I didn't know where the answers to his questions came from, but I must have made them up. Beth was furious. She felt that he had no right to hypnotize me without telling me what he was doing and without asking my permission. She was right. I didn't even know I had been hypnotized, and didn't really believe it then. So my first hypnosis session (a very light trance), something I'd agonized over doing for months, wasn't as scary as I had thought it would be. I'd read that abductees were usually easy to hypnotize. I was convinced I probably couldn't be, yet I never realized it could be that simple.

I left the conference at M.I.T. with more knowledge to file away and sort out later, yet still unsure of my own role in all this abduction business. I still felt I wasn't really involved, maybe peripherally in some way (as a caretaker of Beth?), yet with more questions than answers, despite the information overload. I took one extra thing home with me—fear. It had been wonderful to be able to talk openly about abductions, gather information, and feel needed, protected and accepted. But all that would change when I got home again. It was back to that secret life I led where I had few people to talk with about the most important discovery of my life.

I was afraid people would think I was crazy, and above all I had led my life as a person who wasn't weird. People thought highly of me, I have done well in my career, yet now I was beginning to think about things that only weirdos and freaks talked about. Tears of fear were not the reaction I expected after the conference. But, by the time we arrived home, I knew I would have my shell of normalcy firmly in place.

The drive home seemed uneventful; we were both exhausted. It wasn't until several weeks later that I would finally get up the nerve to confront Beth with the fact that it took fourteen hours to get home. We had made the trip to Boston in nine hours.

29. Beth

I had to admit that the conference was not what I had expected. I had been so ignorant of the phenomenon and its history that every little bit of related material that came to my attention affected me as strongly as hearing of a flock of aliens waiting to be interviewed would have affected the more informed participants. This ignorance on the subject of UFOs and alien abductions had been deliberate—or at least preferred, for me. I felt that the less I knew, the less chance there would be of confabulation; both to myself and others involved in Anna's and my situation.

Anna, on the other hand, felt quite differently about it. To her, the more knowledge she had, the easier it was to cope, to maintain control over her life and environment. Anna needed to know how and why things happened. This strategy had worked for her most of her life, but I feared it would not help much this time. If she were to accumulate every ounce of knowledge about this phenomenon, she would have no more control over it than the rest of us!

Still, the atmosphere at the conference was certainly stimulating. I was particularly impressed by the genuine concern expressed by both the abductees (and contactees), as well as the therapists and those wonderfully unobtrusive "interested observers."

Although the Abductee Panel (which consisted of Anna and myself along with six others) had been traumatic for all of us, we got through it and perhaps were even strengthened by the experience. Our candid expressions of helplessness and fear were exposed, yet somehow diminished, by the telling. This healing effect would prove beneficial to me in helping manage my emotions and renew my determination to resist this intrusion. As difficult as the panel discussion had been, I actually felt better for having participated and realized Anna had been right in pressing me to attend.

A little overwhelmed, but still electrified by the day's events, we drove back to our motel that evening in high spirits. Mass Pike was nearly deserted as we sped west away from the city, bathed in the yellowish glow of the arc-sodium lights which lined both sides of the turnpike. Suddenly the row of lights on both sides of the thruway blinked out in unison as we drove past, flashing on again behind us as if the car had tripped invisible power lines as it rolled by. I watched through the passenger's side-view mirror as the lights flicked on like magic, glancing at Anna to see if she'd noticed this eerie phenomenon. She didn't appear to have seen, or she hadn't considered it worth commenting on, and being unable to let it go without mention, I told her what I'd seen and asked if she didn't think that was rather odd.

"Oh," she mumbled, "That happens all the time."

It happens all the time? What did that mean?

"It's not unusual," she added after prompting. "When Nancy and I are driving home from work after dark, the lights usually go off whenever we pass them, then they come back on again after we've gone by."

She said this so matter-of-factly I almost believed that this might be a common occurrence for most people, and that perhaps I'd just never noticed it. But I knew that wasn't so. I certainly would have noted such a thing happening—and would have remembered it, too. I told Anna I hadn't been aware of it before and asked if she knew of other people who had reported having this effect on highway lights.

No, she admitted. But she believed it had happened to her often enough so that she no longer paid much attention to it. "There must be a lot of people out there who interfere with electrical currents; maybe it has something to do with magnetism, a kind of static electricity that interrupts the flow."

I'd heard of electromagnetic (EM) effects during the conference, but didn't really understand what this had to do with the UFO or alien abduction phenomenon. Perhaps this was what Anna was experiencing. I promised myself I'd ask others if they were able to affect electrical current in a similar manner.

We reached the motel without further incident and collapsed into bed, thoroughly exhausted. As it was unusually chilly for the middle of June, I had kept my bed quilt for added warmth rather than turning on the room's heater, too tired to fool with the buttons and settings. Normally I am uncomfortable with the weight of a heavy quilt or spread and prefer to sleep cool, but on this night—after such a disturbing day—I felt the need to wrap up in something warm and cozy.

When Anna and I awoke at about 6:30 the following morning, we were surprised to find I had three pillows on the head of my double bed while she had only one—on the foot of her bed. We couldn't figure out how this had happened! My quilt (the satin-backed variety), which I had made a point of keeping on upon retiring, was neatly folded at the end of the bed. On previous nights, not needing or wanting the quilt over me, I had pushed it to the end of the bed, finding it lumped on the floor the next morning. The satin backing would not permit the quilt to stay put during the night, no matter how quietly I slept. I couldn't imagine having had the energy or wherewithal to fold down the quilt sometime during the night or, having actually done that, managed not to disturb it from its resting place!

These details may have seemed trivial under normal circumstances, and indeed we both admitted that we were probably overreacting, but with so many other "seemingly trivial" events indicating an abduction-related incident, Anna and I felt they were worth noting.

As had become a part of our normal morning routine, we both checked our bodies for any unusual, recently developed marks. I was dismayed (yet not surprised) to find a large bruise on my left leg and a crescent-shaped scrape on the inside of my left lower arm. These marks were not there the previous night.

(Later, Anna discovered a small puncture on the inside of her lower arm, though we hadn't spotted it that morning.)

The following day, as Anna and I were seated in the upper row of the gallery listening to one of the morning's lectures, I began to feel "separated" from everything, as if I were drifting off to sleep, though I was fully alert and could plainly hear the speaker. The room became very quiet; not as if the normal sounds of the crowded lecture hall had abruptly ceased, but rather as if I had suddenly gone totally deaf. I saw the speaker's mouth move, but couldn't hear a word he was saying; I saw people in the audience shuffling paper, taking notes, stifling coughs and sneezes, fidgeting and whispering, but all this went on without my being able to hear any of it.

Before I could contemplate the reason for this mysterious loss of hearing, I suddenly found myself on a landing beneath a large sun-filled window. The dark blue tiled floor was shiny, as if recently waxed and buffed, and reflected the light pouring in from the window to my right. I wasn't sure where I was, but suspected I might be in one of the buildings somewhere on the M.I.T. campus. I felt the presence of someone else, but couldn't see who or where this other person was. This "someone" demanded that I "go back now," that I was not supposed to be there. I resisted, saying that I didn't want to go back. The command was repeated, more urgently, and I finally agreed to go, although I had no idea how I would accomplish that feat. I didn't even know how I had gotten there!

I did, however, succeed in returning to the lecture hall, where I saw myself sitting next to Anna. It was so strange viewing myself that way, from above and to one side, yet feeling quite whole and unique, as if this other me was someone else entirely. I was sure everyone could see me, yet no one looked up—not even Anna. I was omnipotent! I could go anywhere, see everything. The world was my fish bowl! But then I felt myself falling, dropping into my physical body and being consumed by it as if it were a feather mattress absorbing the weight of a huge boulder.

That free, airy feeling I barely had time to appreciate was gone, and I was immediately bombarded by the noises within the crowded room. Every shuffle, sniffle, grunt and hiccup sounded too loud. The noise was painful and I covered my ears. Eventually the din settled into a bearable range, the speaker's voice dwindling to slightly less than a roar. Feeling oddly sluggish and bloated, as if I'd swallowed a whale, I nudged Anna and told her something very strange had happened to me. After briefly describing my experience, Anna admitted being unaware that anything had happened. She had no feelings of being watched (though we both had felt uneasy earlier that morning) and did not experience sudden "deafness" during the lecture as I had.

I still felt uncomfortable talking openly about my experiences, except with a very few of the people we'd met at the conference. I had been taken aback (and admittedly annoyed) after the Abductee Panel's session several days before when Budd Hopkins, the author of *Intruders* and *Missing Time,* approached me before I could escape from the lecture hall. I was in no condition to talk to anyone and felt if I didn't get out of that room I'd explode! I was pinned between Budd and a blackboard which had been pushed against the front wall, trying without success to slip out from between them before I fell completely apart. My emotional talk was still fresh in my mind and heart and I desperately needed a change of scenery—and mood.

As Budd talked, and I replied with barely polite noises, he finally sensed my anxious state and apologized for intruding (no pun intended). I promised I'd elaborate on my family's experiences at a later time, after I'd had time to collect my thoughts. He understood and was good enough to agree, but I secretly feared our next meeting. This man was a famous writer, investigator and hypnotist. Surely he could not be interested in anything I might have to report. After all, I had so few memories, knew very little about the phenomenon, and wasn't even sure how much I believed about my own experiences, let alone anyone else's.

There had also been a great deal of contention during the discussion periods following talks about whether undue influences were being brought to bear on experiencers by the beliefs of the hypnotists. It was implied that some hypnotists and investigators imposed their beliefs that the aliens were either good or evil, therefore influencing their subjects to view the aliens in the same light, whether or not the subject's conscious memories reflected it. It was, in fact, a divided camp. Many of the investigators who were actively employing hypnosis to retrieve repressed memories within the investigative process had definite opinions on the alien's motives, yet all denied imposing their beliefs on their subjects. One such accuser went so far as to imply that all the subjects of certain hypnotists always viewed the aliens as either good or evil, depending on which biased hypnotist was involved.

Another questioner posed this theory: that although it did seem that experiencers working with certain investigators tended to fall into either the good or evil alien group, that did not necessarily mean they were being influenced by the hypnotist, since all agreed the hypnotists were experienced and professional in their approaches. Perhaps, the speaker suggested, there were more—and different—races of alien beings with separate motives. Some may indeed be friendly and considerate, others abusive and threatening. This controversy raged on throughout the conference and was never comfortably, or amicably, resolved.

The question of whether the aliens are good or evil had never really concerned me before this controversy came to light. I had not realized there might really be friendly aliens who didn't frighten the crap out of their victims! I knew the contactees felt that the aliens were their "Space Brothers," but I had seen this as a harmless (and probably imaginary) coping mechanism for people who could not otherwise contend with the fear and trauma from the intrusions into their realities. Now I wasn't so sure. What if there *were* other alien races? If these creatures did exist—and now it seemed they actually might—why wouldn't there be more than one race? Humans certainly came in a huge variety of races, creeds, colors and beliefs; in addition, many within the same race, creed, color and belief structure held opposing political convictions. If human beings were so diverse, why not alien beings? Would that make them less alien?

Probably not.

Despite my apprehension over talking again with Budd Hopkins, I was glad we had found the opportunity later. He was sensitive and kind, and assured me he was not trying to coerce me into revealing anything I was not comfortable talking about. I relaxed, finding him easy to talk to and sympathetic to my confusion over the validity of my memories. I was surprised when Budd invited Anna and I to New York so that we could spend more time together—even try hypnosis if we wished. I suspected Budd wanted more information on the possibility of family involvement and perhaps genetic manipulation within family lines, but since both Anna and I had concluded we had to know more about our possible link to this phenomenon, we happily accepted the invitation.

This could be the solution to our continuing problem of finding an investigator/therapist/hypnotist all wrapped up in one. We were aware that Budd Hopkins was not a licensed therapist; he never claimed to be. That didn't matter at the moment. The conference had broadened our horizons considerably and our priorities had changed from a concern over sanity to a need to know more about our experiences. If it turned out we were not involved, then we could eliminate one possible cause of our distress and move on to other measures.

Therapy could wait. If we were lucky (or unlucky, depending on our states of mind at the time), we might never need therapy.

Amazing. For mature, educated women-of-the-world, we were incredibly naive.

30. Anna

During the conference, Beth and I met a journalist writing an article for the *New Yorker* magazine, C. D. B. Bryan (Courty). We had been told that a few reporters had been invited, but were assured that we did not have to speak with them unless we wished.

We met Courty the first day of the conference. He immediately introduced himself as a journalist, even knowing that I could read his name tag that identified him as a reporter, and asked if we would speak with him. We declined. During the conference he interviewed several experiencers, investigators and therapists for his article. Beth and I shied away from him. We wanted absolutely no publicity, not even a hint of the possibility of someone recognizing us.

Cameras and tape recorders were banned during the conference, yet occasionally, there was a furor over someone who had forgotten. As we became more comfortable with the people at the conference, and with our own part in it, we began to talk more with the therapists and investigators. Beth even agreed to talk with Courty one day at lunch, after he had agreed to keep any information we gave him anonymous. He was intrigued by our story and kept flattering us that our panel discussion had been the highlight of the conference for him. He said that we were real to him, that we had come across as the most believable of the abductees. It was probably because our terror was so apparent and we were so new at trying to accept what was happening to us.

Beth had read articles by Courty and thought his writing was cynical, skeptical at best. I'd never heard of him. I hadn't read the *New Yorker* in years. I was surprised when she agreed to meet with him again on the afternoon the conference was over. He openly admitted to being a skeptic about this UFO stuff, yet he wanted to understand why we believed it was true, or at least be able to come up with an alternate explanation. Beth was all for that. I felt that if we could convince him of our reality, he would be able to write an article, for a national magazine, that might allow others to begin to accept it, and us, as well. He seemed genuinely interested in our experiences, and in us as people. He felt that this was a very human story, that the effects on the abductees were the most important aspect of the conference.

After the conference, we ate lunch and then went to his hotel room for the in-depth interview. Courty asked if he could tape the interview. We agreed on the condition that he keep our names and the location of the farm secret. He agreed.

Beth did most of the talking; after all it was her story, I was just an interested bystander who had a few things in my past I hadn't yet been able to explain. She related details of the incidents she had talked about in the panel discussion and other incidents she had not discussed. She started with the September 1989 UFO lights over the barn (I chimed in with what I could remember seeing); the first missing time incident on December 15, 1991; the second missing time episode on January 2, 1992 that prompted me to get help from The Fund; and continued with these:

Jan. 12   Beth woke up in the morning, while in Pennsylvania, with a deep triangular burn on her right hand. A nurse suggested it looked like an applied chemical burn; a doctor, who saw it a week later when it had black edges, diagnosed it as a laser burn.

Feb. 10   Beth awoke in the middle of the night feeling like someone was pulling her out of bed. The next day she had bruises and fingerprints on her leg and a bruise on the back of her right hand with three holes in it.

Mar. 8   Beth's missing time (1 hour) with vehicle interference (the electrical system) and the naked "man" in the huge cowboy hat.

Mar. 13   My bedroom visitation by the huge man, with attendant electrical failure only in our bedrooms.

May 17   Our shared dream of little greys in the bedrooms, and shared pregnancy symptoms.

We also discussed some earlier incidents that we thought were somehow connected:

Beth's incident with her father when she was ten years old when their car stalled and they arrived two hours late for her father's business meeting;

My two incidents (in 1974 and 1991) when I was blinded for a few days from wearing my contact lenses;

Other incidents of electrical interference including street lights, copying machines and computers; power failures in our bedrooms (always between 3-4 A.M.); and lights and televisions turning themselves on and off—even when unplugged.

After listening to Beth relate some of her fears to Courty, I found myself more at ease with him and told him some of mine. When I related my deep-seated fear of always being different from the rest of my family, my grandmother's refusal to accept me as a child of my father, my scarcity of childhood memories, and the fingerprint bruises I'd find on my arms, he offered an alternative explanation: child abuse. He'd just finished researching a book on child abuse, and indicated that what I was describing could fit that pattern. Mind you, he didn't suggest that my parents were the abusers, he wondered if I thought that the aliens might have done it. I'd so wanted a different explanation, but I didn't like this one any better. I dismissed it out of hand. I really didn't think I was a victim of child abuse.

We took a dinner break and ate sandwiches while sitting next to the Charles River. As we watched the lights of incoming planes and the occasional helicopter, we discussed how to tell normal lights from UFOs. I find comfort in a saying that our friend Sue came up with: "If it looks like a plane and acts like a plane, then it is a plane." It's something I rely on frequently. We went back to the hotel in a more relaxed frame of mind to continue the interview.

I told Courty about Beth's and my pregnancy symptoms (after a few drinks). He was very matter-of-fact about it. I'd tried, unsuccessfully, to talk to one of the investigators at the conference who was an emergency room doctor. All he told me was that women did come to the emergency room with strange pregnancies (actually the lack thereof), but I felt he thought that it was all in their heads. A typical male reaction: Hysterical woman thought she was pregnant and now she's not, aliens took it. I was not impressed by his attitude, especially from someone who investigates abduction phenomena. He said that he'd never seen any proof of a missing fetus. What's proof? It's gone, isn't it?

Anyway, Courty asked the obvious question: "Have you had sexual intercourse with a man these past few months?" Alas, no. It had been at least a year since I had been intimate with a man, and I had no energy to devote to developing a relationship lately. Besides, how could I develop a relationship if I was an abductee? How do you tell someone and not have them think you're crazy? "Sorry, dear, but don't worry if you wake in the middle of the night and see me floating out the wall, or you hear or see strange things when you're with me—it's just the aliens!"

But I have erotic dreams. To compensate? One happened a few days before I began feeling pregnant. What was strange about that dream, thinking back, was that we had proceeded all the way to intercourse, I had an orgasm, and was very contented. Most of my erotic dreams never progress that far; I just feel really good and loved. I couldn't explain why my vulva was sore the next morning; dreams usually don't cause physical symptoms. Usually along with the erotic dreams, I feel particularly horny. My hormones are flying high for a few months, for no apparent reason, and I feel like attacking any man in sight. I'm too young for menopause. Maybe I'm just a horny older lady, with the hormones of a twenty year-old. Beth admits to having the same thing happen to her; maybe it's just normal. Sometimes it just feels good to be stared at and appreciated by men wherever I go.

We ended the interview with discussions about coping mechanisms. Beth and I agreed that having each other, sharing the misery as it were, was the best way for us to cope. It was the only way. We were luckier than most experiencers we met at the conference. Many had no spouse or friend to talk with. For better or worse, we were in this together.

31. Beth

"I can't believe we're doing this!" I said to Anna as we followed Courty Bryan to his hotel room across the bridge from the campus. We found a parking place across the street from the hotel, neither of us moving to get out of the car.

Anna sighed, apparently annoyed by my skepticism. "I trust Courty," she stated firmly. "If you're uncomfortable with this, you don't have to say anything."

I wasn't sure whether she was angry, or as nervous as I was and unwilling to admit it. I trusted the man, as much as I trusted anyone at that moment, but I was still worried that we were about to do something we may regret later.

C. D. B. Bryan, better known as Courty, was a reporter for the *New Yorker* magazine, and one of the few invited from the press to witness the conference proceedings. We were informed that Dr. John Mack had absolute faith in Courty's professional integrity, and had assured the experiencers that the reporter would treat them respectfully and not record, print or otherwise divulge any information he learned from the experiencers without their expressed consent. This was high praise, coming as it did from one of the most respected people at the conference, and it paved the way for Courty to arrange for a number of interviews during breaks in the conference proceedings.

Anna and I had both talked with Courty briefly, permitting him to make some notes for his planned article in the *New Yorker*, but we kept our comments generic and were not inclined to go on tape. Tape recorders had not been permitted during the proceedings, so whenever there was a recess and one or more of the experiencers slipped outside to sit under the trees, Courty could usually be found nearby, pen and pad in one hand, recorder case in the other.

Courty did not push, but his presence around us became so commonplace that we began to see him as one of us, no longer feeling quite so reticent about speaking openly about our feelings and experiences. After the Abductee Panel and its attendant revelations, Courty seemed to spend more time with Anna and me, and for longer periods. Even though we'd declined an interview, neither of us felt uncomfortable talking with him. It was inevitable, with this more relaxed relationship developing, that we would also relax our taboo on agreeing to a formal interview. Arrangements were made to have that interview on June 17, at Courty's hotel suite in town.

We were all a little stiff as we waited for the elevator in the hotel lobby, each of us entertaining our own thoughts and concerns, no doubt. Once inside the suite, Anna and I seated ourselves on the sofa and made small talk with Courty who was busily preparing drinks for us from a tiny kitchen off the entranceway. Sipping at our drinks and trying to keep the conversation neutral, we watched as Courty set up his recording gear and arranged the microphone so it would pick up our voices with the least interference. A newly unwrapped tape was placed in the machine and the recorder and sound were tested. All seemed to be in order, but now that the equipment was running, we didn't know how to begin.

Courty was an expert, though, and he began by asking us individually to tell him about ourselves; nothing very specific, just a little background to help us relax and get comfortable with the process. I was elected to start, and even though I had no fear of microphones and was not particularly nervous about answering Courty's questions, I was concerned that I wouldn't be able to contribute much to his article, and that what I did say would come out sounding crazy to his readership. Courty had already expressed his skepticism of the alien abduction phenomenon, but readily admitted that he was finding it more and more difficult to maintain that cynicism in view of what he'd heard from us and the other panel members during the conference. This spontaneous admission helped me to relax at last, and I concentrated on answering his questions instead of worrying that we might be portrayed as weirdos in his article.

I had begun relating a recent abduction memory, scanty as it was, when I abruptly slipped out of the story line and into what felt like a flashback. I saw one of our dogs, Cricket, tearing through the house, whining and crying in abject terror. Her tail was tucked between her legs as she rushed past me (I was standing in the dining room) into the kitchen then disappeared down the stairs leading to the basement. The flashback faded and I sat frozen on the sofa, my fingers gripping the edge of the cushion as if I were afraid of falling off. Seeing that I was quite upset by this memory, Courty shut off the recorder immediately. He apologized for not breaking off the session sooner, but I didn't feel that would have stopped the flashback. Hoping to give me time to collect myself, he left us alone and went to make a pot of coffee. Anna and I consoled each other, trying to figure out when this event might have happened and what it might mean. Suddenly the recorder came on again—yet no one had touched it. No one had even been near it since Courty turned it off. We sat transfixed by the running tape, calling for Courty to come back.

Courty had no explanation for the mystery, but I suspected he believed Anna or I had turned it on unintentionally. We decided to forget about it and continued recording. The mike was directed at Anna this time as she told of her background and certain events that puzzled her. As the interview progressed and Anna began to touch on more troubling events, she also became distraught and the machine was once again turned off to allow her time to collect her thoughts. When she was ready to continue, Courty rewound the tape just enough to remind them of where they were in the interview. When the tape was set to play, no voices were heard. He wound it back again, further this time, and finally we heard sounds emanating from the machine—but it didn't sound like our voices! The entire tape was garbled! It wasn't just hard to understand; it was closer to hearing a foreign language spoken backwards at high speed. Stranger still, only Courty's voice came in loud and clear while Anna's was completely unintelligible.

We thought it would be wise to check the other tapes, though the ones we had used were wrapped in cellophane and only opened as they were needed. When my interview tape was played back, Courty's voice was again distinct while mine sounded faint; Courty had to hold the recorder close to his ear in order to hear my voice. Each time he played it back, my voice became fainter until Courty was unable to pick it up at all. The tape was allowed to play through so we could determine if any of my interview would be discernible. Further along in the recording, my voice began to come in more clearly and louder—but this time we heard hysterical laughter—my laughter. But I couldn't remember laughing like that at any time during the interview! Nothing we'd said or done since arriving at Courty's hotel could be classified as *amusing*.

We took a short break and walked along the river to relax and collect our thoughts. When we returned to the hotel suite, we were interrupted by a young woman representing the *Atlantic Monthly*. She had also been present during the conference, though Anna and I had not met her before. We were introduced and asked if either of us minded her sitting in on the interview. We weren't uncomfortable with her—and besides, at that point in the proceedings, it hardly mattered how many witnesses were there. Nothing seemed to be going according to plan and very little was being recorded successfully or consistently.

We were chatting amiably with the newcomer, filling her in on the taping problems we'd been experiencing, when Courty began complaining that he was having trouble removing the tape from the machine. It appeared to be jammed, but Courty was reluctant to force it and chance ruining what little might be salvageable on the tape. The other reporter decided to give it a try, but had no better luck. Suddenly I was sure that if Anna and I left the room, the tape would free itself. As soon as we had stepped into the alcove, the tape popped free on the first try. No one wanted to venture an opinion on what all this meant. We were all simply relieved that it hadn't been necessary to tear the recorder apart in order to release the tape.

We had about decided to quit while we were ahead when the woman from the *Atlantic Monthly* jumped up in surprise, pointing to the extension cord strung across the floor, still connected to the recorder. We followed her eyes, not understanding what she saw. Then, as if fearing someone might hear, she whispered, "Did you see that?" We shook our heads, wondering what else could possibly happen that would be stranger than those things we'd already witnessed! The cord was entwined around Courty's chair leg across from the sofa (Courty was no longer sitting in the chair), and so far as we could tell, there was nothing odd about it. Suddenly the cord slithered several inches away from the chair leg, then back again, like a snake changing position to take advantage of the sun's warmth. There had been no one near the cord or the tape recorder. I felt cold fingers tickling my spine and the hair on my arms and neck stood at attention. This was truly weird! I looked at the reporter, raising an eyebrow in query, hesitating to speak aloud for fear the cord might do something even stranger.

"Did you see it?" she asked again.

This time I could answer in the positive, though I still didn't believe it. I glanced at Anna and saw that she was calm, seemingly unaffected by this new unexplained event. I asked her if she'd seen what we'd seen.

"I didn't want to look," she said flatly, keeping her eyes averted.

Strangeness overload, I supposed. Had she become numbed by all the strangeness around us? I hoped so. Perhaps Anna had finally found her coping mechanism—oblivion.

If one doesn't look, one can't see. If one can't see, one does not have to explain.

See no evil, hear no evil, speak no evil.

I wanted to be like her.

chapter 7

Exploration

32. Anna

The summer of 1992 kept me very busy trying to shake off the feeling that something was really happening to me. I knew that things kept happening to Beth, and we had many unexplained occurrences around the farm and things happening to the people doing business with the farm. I kept finding small scars and cuts on my body that I hadn't noticed before. But how many people check their body for marks every day? Abductees! It became a daily ritual for several months to check for new marks, and sure enough I'd find them, but I still wasn't convinced they were really new. I'd just missed them the last time I checked.

The stress of it all was getting to me. I still wasn't doing much at work. I used past knowledge and experience to get by. My work standards are very high and I felt guilty about not accomplishing more. I wasn't learning anything new. It was very hard to get enthused about any project I worked on. Besides, no one else seemed to care either. I'd turn in projects weeks or even months late and give some lame excuse. They bought it. I remember one telling day at work when my supervisor said he needed to talk with me. We set up a time to meet later in the day, and I stewed for hours. Now I was really going to get chewed out for not doing enough, isolating myself from co-workers, playing computer games, spending too much time in the library doing research (reading UFO books!), yet still not completing projects. I knew fifty reasons why I deserved to be lectured. After lunch he came by and apologized for leaving me wondering what he wanted to talk about. He must have known I was worried. All he wanted to do was to tell me that I was going to receive an outstanding performance rating!

The stress was also affecting me physically. I got shingles in July, followed by recurring bouts of diarrhea and upset stomach. I had headaches almost all of the time, and I wasn't sleeping well. I went to an internist and she found some diverticulosis, gave me some pills for the headache, and we tried a couple of varieties of antidepressants; none of which helped. I'd also had a CAT scan done; everything normal. You'd think that would have been a relief, but it wasn't. I still had persistent headaches.

I'd also had some radical changes in my eyesight in July and had a hard time getting the correct prescription. This was the last straw for me. Beth had had similar changes in her eyesight before I did. She was convinced that I was being abducted with her and that the aliens had changed our eyes in some way. I even brought her a peace offering the day I got the news from the optometrist—a whole pound of dark chocolate nonpareils. I was getting convinced. I decided to see a hypnotherapist to try and find out what was really going on.

We only had one hypnotherapist in our area who had worked with abductees, but he had decided that he would only treat the trauma associated with abductions. He no longer wanted the responsibility of uncovering memories through hypnosis. I'd even made an appointment to see him in September, but then cancelled at the last minute. I wasn't ready yet. Later, I purposefully chose a woman who had no knowledge of abductions. I'd read a lot of the skeptics' arguments and some of them seemed to suggest that investigators and therapists who knew about abductions were implanting these memories in their subjects. As if this was any easier to believe than that these experiences were real! Well, I wasn't taking any chances.

The psychologist I met with had been doing hypnotherapy for almost all of her twenty years of practice, dealing mainly with disturbed adolescents and child abuse victims. She had also written several articles on the therapy she had done with people with multiple personalities. The only thing on abductions she had read was the results of the Roper poll that had been sent to thousands of therapists, hoping to enlist their support to help abductees. I had met her socially at the end of September. I knew her to be open-minded, but that first session where I admitted that I thought I was being abducted by aliens was tough for both of us. She told me that she was retiring within a year and did not plan on taking on any new patients, but would be willing to help me discover the source of my headaches. She frankly admitted that she didn't believe in alien abductions and had never run across it in her practice.

In our first session, we did not use hypnosis. We discussed my background, childhood memories (including the rape), and what led me to believe such an incredible thing as abductions. As we discussed my memories of childhood, I was astounded to find that I really didn't have very many. I thought that most people had lots of memories of early childhood and adolescence. My earliest memory was when I was about two and a half years old when my father was taking care of my sister and me because my mother was in the hospital having my brother. I had just one brief image of eating beans for dinner one night. The next memory I had was walking into my first grade class (I guess I didn't start at this school at the beginning of the year) and having the teacher tell me I was too tall to be in the first grade and that I really should be in the class down the hall! I cried. I had some memories of climbing trees and going to the beach in Norfolk, Virginia, some of living in England and camping in Europe, and a few of junior high school in Florida, but I had very few memories up until the time of high school—bits and pieces here and there, but no real time line of memories.

I think most of the memories I had existed because I had seen pictures of the family at different points in time. My father was an avid photographer, but I had a hard time connecting the little blonde girl in the pictures with me. The therapist explained to me that this is quite common in children who have been abused and is consistent with the development of multiple personalities. I assured her that I hadn't been abused. I'd have remembered it, wouldn't I? No, she explained, that was just the point. Many people don't remember. As far as the periods of missing time that I couldn't account for, people with multiple personalities often have time lapses where another of the personalities has taken over. Maybe that was the reason I was so intrigued by the accounts of people with multiple personalities. Maybe I was one (or many, as the case may be) and didn't know it. Now I'd have an explanation for all this strange stuff that was more acceptable to the rest of the world. I had my out!

We scheduled a hypnosis session for the morning of October 16. Beth went with me. The therapist again admitted to a healthy skepticism about alien abductions, explained that she didn't normally do hypnosis this soon with new patients, but was willing to help me to get rid of the debilitating headaches if she could.

She also admitted that she was looking for a more mundane explanation than the little gray guys. I asked that Beth be allowed to sit in. It wasn't something she usually did, but she allowed it. Since I did not expect to have anything useful turn up with the hypnosis, I did not bring a tape recorder. The following account is from the notes in my journal.

She hypnotized me and took me back to when the headaches first started to try to discover what caused them. We went back to the middle of July of that year. It was hard for me. I never really thought I was hypnotized and had to keep telling myself I was. I saw lots of colors, mainly red at this point—sort of misty, but with more substance to it. When the therapist asked me to go back to the incident when the headaches started, I felt that I fumbled around for a long time being nowhere, my vision was filled with black or red fogs. Beth says that I responded immediately to the request by jerking in the chair (they both flinched), got very upset, was crying hysterically and kept screaming about the eyes, *the eyes!* It was the big alien eyes, filling my entire vision. The therapist calmed me down and asked me to describe them—big, dark, staring at me from less than two inches away. I even covered my eyes with my hands to make them go away. No luck. We should have brought the tape recorder! The rest of this is not in order of occurrence—just what I saw during the hour I was hypnotized.

I eventually saw the whole alien being and was surprised at how thin and tall he was. The ones I'd read about were less than five feet tall. This one was at least five feet six inches tall. The therapist asked what the alien's name was—I told her Joe (or Jo; I don't know the spelling.) The only Joe I knew was a guy I dated for about five years, six years ago—this was definitely not him. It surprised me that the alien had a name and that he was familiar to me. Each time he came into my vision, the red mist changed to a dark, brilliant, almost electric blue. Every time I saw that blue color under hypnosis, I got very frightened and cried. (Beth mentioned afterwards that I don't wear any blue clothes.)

At another point I saw the tops of four or five alien heads with just the edge of their huge eyes showing. The therapist asked me if I was standing up, and I rationalized and analyzed (and said) that if I was standing up they would have to be behind me, floating upside down in the air with just the tops of their heads visible to me, so I guessed I was lying down on a white table. These were the smaller aliens. At this time I saw a huge two- or three-sided "arrowhead," maybe on a stick—maybe not—on my left. It was about four inches long and black. She asked me if they touched me with it, but I didn't think so.

The headache pain came and went during the session. At one time it was so intense I held my head and couldn't see anything. She asked me if they had touched my head. I thought they had, but couldn't see them doing that. The therapist asked me to describe where I was. The walls were all white, soft, marshmallowy. I was in a pie-shaped room with the large end of the slice at my feet. There was a soft whiteness to everything, light came from nowhere yet was everywhere. As the slice went by me it went to a point behind my head. Those walls were straight and didn't have as soft a feeling. The therapist asked if the little guys left when I couldn't see them any-more. No, I knew they were still in the room, but they were farther behind me, not peering at the top of my head anymore. She asked me what I was wearing: jeans and maybe a yellow T-shirt. I didn't feel naked and wondered at the time, wasn't I supposed to be naked? In all of the stories I'd read the abductees were naked when they were on a table. Maybe I made this up. Seemed awfully real at the time I was reliving it under hypnosis.

I got pretty upset describing these things, so Beth says the therapist asked me to return to the beach where I had started the hypnosis session. We'd agreed before entering the trance that I would choose a safe, restful place where I could return if I felt anxious during the session. I had chosen a beach that I used to visit when I lived in the islands. I could clearly picture the soft swells, the warm sun, and hear the waves hitting the beach. When she asked me to go back to the beach now, she described the sun, the waves, the soft billowy clouds casting shadows on the sand. That was it for me. My beach hadn't had any shadows, but I saw shadows in the palm trees behind me and then I knew that there was something very frightening waiting for me in those shadows. I couldn't return to that beach during the session, even though she asked me to go there to calm down several times. I now suspect that I was abducted from that beach.

Since the therapist was not an investigator, she did not know the questions to ask me to find out what else had happened during this abduction, and I didn't volunteer any information. I was terrified of the gray beings, and wanted to leave their presence as soon as possible. I also saw a huge ocean liner coming at me bow first—I felt like I was below it, maybe in the water. It was all black on the bottom of the prow and white above. I kept seeing the red color and sometimes the blue.

At one point I saw myself in a long tunnel with a small green light at the end of it. The tunnel had soft sides. I was thrilled, awed and amazed that I was not walking down this tunnel—I was just moving down it. Floating! I even wiggled my feet at that point. She never asked me to go to the end of the tunnel, so I don't know where, or to what it led. Then I saw the shadow of Alfred Hitchcock! When the hour was almost up and when I didn't have the headache, the therapist tried to bring me back to the beach, and I became very upset, crying. I didn't want to go back.

So she let me remain hypnotized a while longer. That's when I saw other forms in the lights, or blackness. They seemed like amorphous fetuses at times—they were never very clear. At one time I also saw an alien chin—very large and pointed, no lips, no nose, just a chin. Just before she brought me back I saw an eagle. I saw just its head and shoulders. I was in a large clearing, surrounded by tall oak trees, but I was suspended halfway up their trunks. I wasn't looking at them from the ground; the eagle was level with me. I saw a blue open space above the trees.

She eventually brought me out of hypnosis. I felt dizzy, but okay, but still had a slight headache even though I'd taken a pain killer that morning about three hours before the session. I had to go to the bathroom. When I was gone, the therapist admitted to Beth she had never run across anything like this in her twenty years of doing hypnosis. None of it fit any of the patterns she thought she'd find—child abuse, someone hitting me on the head, sexual abuse (she expected to hear the name of my rapist when she asked me if the person had a name, but it wasn't my father's name), multiple personality—but it wasn't anything she'd ever dealt with. She admitted to being frightened by the session, and at a loss—she wanted to help me, but didn't know how. Oh well, so much for normal explanations—if it was something she had dealt with before it would be curable—the little gray guys aren't! I had finally begun to acknowledge my involvement with the aliens.

Beth and I talked about the hypnosis session while driving home and through lunch—I couldn't eat after I'd ordered. We were both exhausted. Beth more so than I. She started having a headache when mine got so bad under hypnosis. Sympathy pains?

I gained no new memories in the ensuing weeks, even though it is not uncommon for other memories to begin surfacing after an hypnosis session. The headaches continued unabated, in fact they seemed worse. I was being awakened by the headaches at about 6:00 A.M., on those days I didn't get up at 5:00 A.M. to go to work, just to take something for them. I was taking up to twelve ibuprofen a day just to keep the headaches at a manageable level.

As the weeks passed, and the reality of what I'd experienced under hypnosis faded, no new memories surfaced and I again began to doubt my involvement. I tried to convince myself that nothing had happened since July.

33. Anna

By this time I had come to depend on my weekly phone calls to Richard Hall. After the M.I.T. conference, I felt much more at ease talking with him; he had become a friend, not just an investigator. Without his support, and the growing interest and support of Bob Huff, who we'd met at Rob Swiatek's, I would now be a physical and emotional wreck. Richard gave me the intellectual support I needed and Bob gave me things to do with my knowledge.

Bob was convinced we could stop the aliens from coming, or at least make it harder for them to abduct Beth. Failing that, maybe we could get some hard evidence of their presence; eye witnesses, photos, video tapes.

Every other Saturday Bob would appear with some new contraption to try out—Bob's toys. It started rather simply. He installed an infrared sensor over Beth's bed to detect movement in her room. The sensor activated a buzzer and a light in my room. Of course we set it so that it wouldn't go off every time Beth turned over in bed. He and I waited for her next abduction so that I could wake up and confirm it, maybe even scare them away, if that were possible.

Needless to say it didn't work for a variety of reasons. Beth's abductions continued and I slept right through them! The buzzer didn't go off (Bob and I assumed that she forgot to set it or had turned it off at the beginning of an abduction) and when I would wake up in the morning with the sensor light on, I didn't notice it. Bob got me a beeper that sounded whenever it detected light, but for some reason I would forget to set it on the night of an abduction. I was getting as bad as Beth at playing into the gray shits' hands.

We then set up a video camera, but that didn't help us get our evidence. It never worked when an abduction took place, or if it did, the tape was blank. All the other nights we had wonderful scenes of Beth sleeping. I should have known better. I had talked with another abductee and she had tried the same thing. One night, when she wasn't made to turn off the camera by the gray shits, she probably got some footage. Unfortunately, she removed the tape from the camera the next morning, destroyed the tape with a hammer, and tossed it in the trash before she realized what she was doing. Mind control before, during and after an abduction seems to be their specialty.

Beth's attitude didn't help either. I was furious with her the day I found out that she had ripped all of the surveillance equipment out of her room. Didn't she realize that all we were trying to do was get some evidence that these guys did what she said they did? Then maybe we could get some protection. She didn't understand my reaction; I didn't understand hers. Oh well, truce time. It was her bedroom, so Bob kept up surveillance less obtrusively.

The abduction nights when he was keeping watch from the guest bedroom, he overslept his appointed rounds and woke up after they had brought Beth back to her bedroom. I think he was switched off by the aliens, not an uncommon occurrence for spouses of abductees, but he refuses to believe that they could do that to him. The first time this happened, he and I were both awakened by the slamming of Beth's bedroom door. The door frame was pulled away from the wall in places! We found her terrorized, with no memories, but with fresh bruises on her legs.

Bob still comes to keep watch a couple of times a month, when he thinks he has figured out a pattern to the abductions, and sometimes he guesses right. We still don't have any evidence that someone else would believe (bruises could be self-inflicted, terror and nausea could be from nightmares or stress), but maybe one day we will.

34. Beth

Good God! When Anna asked me to go up to my bedroom to bring her a book I'd read which she wanted to borrow, I never expected that!

As I walked casually down the hallway toward my bedroom, I did notice that my door was pulled to, but I hadn't expected that I should be alert for something unusual. I pushed the door open then recoiled, my heart pounding, as a shrill alarm sounded. I had no idea what it was, where it had come from, or why an alarm should be going off at all! Before I had time to recover my composure, Anna and Bob, who had come by earlier to check on us, were standing behind me, grinning fiendishly.

"What in the world was that!" I demanded, suspecting a prank, as Bob pushed past to deactivate the alarm. My ears popped unexpectedly. Suddenly it seemed very quiet.

Anna had to work hard to cover her amusement, but managed to explain with minimal decorum, "It's a motion sensor. Bob installed it a little while ago."

My desk lamp was on, though I was certain it had not been on before. Anna saw it too, adding that when the sensor detected motion anywhere near my bedroom door, it activated an alarm on the night stand. In turn, my desk lamp (which was fitted with a special bulb fixture) was plugged into another contraption which turned the lamp on automatically within seconds.

I entered the room cautiously, not yet satisfied that the blaring alarm would not start up again. Bob and Anna stood proudly under the sensor—an innocuous looking plastic globe suspended above the window—as if expecting applause for such a creative and witty surprise. I was still unnerved, still undecided if I wanted that thing going off every time I walked into my own room! Bob assured me that the sensor could not activate the alarm unless the alarm was properly set, and only then if the room was dark. I believed him, thinking that if anyone knew about electronic gadgets, it would be Bob. Besides, maybe it would help. Even if we didn't catch the gray shits in the act, we might at least scare them off—or give them something else to think about!

For the next several weeks the sensor and I lived in reasonable harmony; it didn't bother me and I didn't interfere with it. The alarm-setting routine became part of my bedtime ritual, and although it never did "catch the gray shits in the act," it did provide an occasional scare, keeping me from becoming too complacent.

One evening about 10:00, as I was sitting up in bed reading—a habit I found both enjoyable and relaxing—I felt the hairs stand up on my arms and was immediately alert. The room felt chilly, the air electrified. I waited breathlessly for something to happen, though I don't know what I expected. Suddenly the desk lamp snapped on. I hadn't connected the alarm yet, but since the light fixture worked independently from the alarm and directly in response to the motion detector, I was saved from that dreaded racket. The light stayed on for the programmed fifteen seconds, then snapped off. I never saw anything; there was no one outside the door and no one else in the room with me. It was a mystery, but probably no more a mystery than what the sensor was installed to detect.

I had shrugged off that unsettling event, figuring that electronics, like magic, could not really be explained or fully understood; it was only as reliable as the electricity that sustained it. And the electrical circuitry in the house had already proved somewhat unreliable. During March of 1992, several odd events took place which I have yet to understand:

I had just turned out the light and was about to drop off to sleep when my small color TV came on by itself! It was not picking up a channel, just fuzz and static. Startled, I jumped out of bed, setting off the alarm, of course, and scaring myself half to death. Before returning to bed, I unplugged both the TV and the alarm, then unplugged the desk lamp and moved it to a chair, its cord hanging lazily over the shade like a sleeping snake. So there! That should guarantee me a decent night's sleep for a change, I thought.

I hadn't bothered understanding why the TV had turned itself on; it was just magic after all. But sometime during the early morning hours another magical manifestation disrupted my sleep. I awoke to find the room flooded with light from an unlikely source. Struggling to focus, I squinted into the glare. The light seemed to be emanating from the desk chair. But that didn't make any sense. Why would the chair be lit up? I sat up in bed and rubbed my eyes, finally able to see clearly enough to identify the desk lamp perched on the edge of the chair. But why was it turned on? Then I remembered: I had put the lamp there earlier that evening after the TV incident—but I had unplugged it, hadn't I? I crawled out of bed in slow motion, as if giving the lamp time to change its mind and behave like a normal, unplugged appliance. Approaching cautiously, deliberately ignoring the electrical cord draped over the lamp's shade, I reached down and snapped off the light. The lamp obediently complied and I was thrust into darkness.

I don't recall groping my way back to bed. It's as if time just stopped with that singular, defiant action of switching off the light. Had these mysterious events been preludes to an abduction? Had I been switched off along with the light? Or was it just more electronic magic?

I mentioned the night's troubles to both Anna and Bob the following day, complaining that the sensor contraptions were either malfunctioning or something far stranger was going on. The lamp was definitely unplugged; I could still see the cord lying over the shade. Unless I imagined the whole incident, I had actually turned off an electrically powered light fixture which was not even connected to a source of power!

Then again, perhaps I did imagine it. A dream. That was probably it. There was so much strangeness my mind was compensating by releasing the attending anxiety through nightmares—very realistic nightmares.

Then, at 4:15 A.M. on March 13, the power in Anna's and my bedrooms went out, though the rest of the house was unaffected. Several rooms are on the same circuit and it would stand to reason that should our bedrooms lose power, other areas of the house would also. This power outage coincided with a visitation from a tall male intruder in Anna's bedroom; she found this strange man standing beside her bed looking down at her. When Anna reached for the light, noticing the clock read a little after 4:00 A.M., she was suddenly overcome with drowsiness (despite the presence of an unknown figure in her room) and dropped off into a sound sleep.

This would not be the last time such electrical anomalies besieged us, and we were to discover over the following months that other abductees had similar experiences with unexplained power outages.

While these electrical problems continued, other odd events contributed to sleep deprivation as well. Dreamlike visions of small gray beings peeking around furniture, floating near the ceiling, and raising me from the bed without the use of force kept me in a constant state of uneasiness. In an effort to both console and rationalize, Bob suggested we might consider more elaborate surveillance devices connected to a backup battery that would preclude interference from a power outage. At the time, I was willing to try *anything* that might permit a decent night's rest! The week before, Bob had donated a rather unusual lamp, called an Illuma Storm, which looked suspiciously like streaks of frenzied lightning trapped inside a glass ball. The Illuma Storm could be programmed to respond to sound or touch (such as the touch of a finger on the glass) or be set to a continuous display. This spirited light show was not altogether unpleasant, but it took a few nights to adjust to its constant flashing in the dark. I found myself wishing for an interruption in power!

Just as I began to ignore the thing, Bob advised keeping it set to the sensory mode, so that should the intruders drop in, it might give silent—yet visible—warning. This seemed logical, yet when I was awakened for several nights running by the Illuma Storm's sudden and sporadic bursts, as if an electrical storm had developed inside my room, I questioned the device's reliability. It seemed faithful to its own obscure urges, no matter what stimuli might have otherwise activated it. Since it was flaring up often enough to disturb me, I decided to leave it in the active mode all night. I would just have to tolerate it.

The parade of "electronic toys" was moving into high gear. Next came a light sensor, which emitted a piercing alarm whenever it detected light in the room. Naturally, it had to be set after the room was in virtual darkness already. (I hate to admit how many times I inadvertently triggered the alarm by setting it while I could still see what I was doing—or forgot to turn it off before the sun rose!) After only a week, I had had enough of this new alarm and turned it over to Anna to play with.

Initially, the installation of more sophisticated equipment was reassuring—surely one of these instruments would detect the presence of intruders in my room. But one by one, toy by toy, I began to feel invaded by the very devices intended to provide me with comfort and peace of mind. My bedroom had begun to look like a Radio Shack warehouse—or an FBI experimental gadget lab. However, their effectiveness against the gray shits was questionable at best.

One early morning, before daylight, I was startled awake, not sure what had disturbed me. I was anxious, but unable to move to

turn the light on. Instead, I stared at the Illuma Storm, only then realizing it was behaving erratically, exploding in short bursts of very
bright light, then winking out leaving a faint afterglow imprinted on
the glass. A final brilliant blaze caused me to close my eyes—then
nothing. I must have fallen back to sleep. When I got up later that
morning, I discovered all the alarms, sensors and lights were unplugged! Had I done that and not remembered? I didn't think so.
What good were all these sensors if I were going to unconsciously
render them inoperable?

I felt guilty that all those gadgets could so easily be disarmed,
since by their warnings Bob's nightly vigils could have been considerably less time-consuming and exhausting. But if I didn't know I
was turning them off, how could I stop myself from doing it? Bob,
not the least discouraged by my early morning single-handed sabo-
tage, recruited backup. Although he had spent many evenings keeping watch over us, sacrificing time with his wife and family, it was
understandable that he felt unable to maintain the vigil unendingly;
there was no telling when the gray shits might come again and Bob
was reluctant to leave us without an observer.

In true investigator fashion, Bob located two young men who
had expressed an interest in observing a few nights a week. Since
there was a comfortable—but tiny—guest room upstairs, they would
be able to keep watch in shifts, never leaving us without a "keeper."
Another volunteer was found to fill the void on other nights, though
this proved to be an inconvenience: The fascination for encountering
the unexplained face-on endures, we learned, about as long as one's
physical stamina—or, as in this case, until the observer becomes a
participant.

On a pleasant July evening as Bob kept watch upstairs and his
friend took up position outside the house, an event took place which
would not be fully investigated until some months later. While this
abduction proceeded, the outside observer was strangely obsessed
with taking a tour of the stable, where he remained engrossed for the
duration of the incident. The following morning, as I recounted what
I had then believed to be a dream, we learned about the observer's
"stable tour." The man seemed confused when questioned about
this, explaining that he'd just felt like taking a walk. *Into a dark barn?*
I asked. *Wasn't this the man who had no interest at all in horses, let alone
stables?* It was, Bob assured me. *And he chose that particular time of the
morning to go on a tour?* I asked the man how long he'd been in the
barn, but he didn't appear to know. Long enough for him, I concluded.

One observer down, three to go. How long would it take for *them* to overload? The two young men had already witnessed unexplained lights (or beams of light) directed at the farm, yet still agreed to observe overnight. Although we were not aware of anything having happened during their vigilance, they too lost interest and faded away. Bob was again holding down the fort alone.

Without other observers to help shoulder the load, Bob had to rely on remote observation. This was accomplished by setting up our video camera so it would begin recording at 2:00 A.M., about the time most of the events took place, and shut off automatically at 4:00 A.M. On the trial run, it was quickly discovered that the cam-corder required more artificial light than was currently available, that being the Illuma Storm. I refused to sleep with an overhead light on just so the camera could focus! Sleeping had become a real challenge over the past month, what with all the other sensors beeping, flashing, whining and squealing! I was spending more time setting alarms than I was sleeping—and that was only if the little gray shits left me alone! I was going to put my foot down on this last one. If a video camera might help us detect something, fine. I was willing to give it a chance. But it would have to depend on the available light source. I had at least become used to that and had learned to sleep with it on.

"Besides," I reminded him glibly, "they've never shown up in the dark. They bring their own light with them."

Nothing ever came of the camera snooping, although I spent several restless nights lying awake and watching the little red "on" light blink rapidly, like a ruby with the hiccups.

By late summer I was inundated with detection devices, wondering if I would ever again be able to go upstairs, close my bedroom door, climb gratefully into bed, and turn out the light—without plugging in plugs, pushing buttons, flicking switches, fumbling in the dark to find my bed, then forcing myself to sleep while machines bleeped and bumped in the night—all this just so I could be awakened in the wee hours of morning by an impossibly bright flashlight held by a groggy Bob as he peeked in on what he surely must have believed was a soundly sleeping potential abductee.

"*I've had it!*" I announced to myself. Taking the stairs two at a time—something I wouldn't have believed I was still capable of doing—I tore into my bedroom and began dismantling every sensor within reach, tucking them into a drawer for the time being. I would not allow myself to be taken over by fear and superstition! If abductions were going to take place anyway—despite all this mechanical wizardry—then there was no point to all of it. I was tired of the bedtime alarm-setting routine, tired of being startled awake in the middle of the night by well-meaning observers with blinding flashlights, tired of having my privacy invaded. I appreciated Bob's efforts and didn't wish to insult him by refusing his well-intentioned help, but it had come down to priorities: his or mine.

Since my mental health was more important to me than whatever research material might be gathered by surveillance, it seemed prudent to dissolve the spy network temporarily. I didn't want to burn all my bridges; I may later agree to reinstate the process, but for now it was more of an invasion than a comfort.

There were a number of benefits to having these sensors, since one never knew which device might actually detect something. Some of the sensors did, in fact, reflect activity in my room which might otherwise have gone unnoticed. The Illuma Storm predictably flared whenever electrical activity increased; the motion detector was sensitive enough to pick up the least disturbance, and when used in conjunction with the special light fixture, could be relied upon to switch on the lamp. When this lamp was in the guest room, Bob could be immediately alerted. As for the other devices, I apparently had no conscious objection to turning them off, effectively neutralizing them.

Accepting this, I decided to allow the Illuma Storm and the motion detector to remain active, though I would later insist that the motion sensor's light fixture only be used in the guest room when an observer was on duty. I didn't want to appear too obstinate! (The Illuma Storm would eventually fall prey to an overdose of electrical stimulus and become essentially powerless. Anna was gifted with her own Illuma Storm, which to this day performs admirably, but perhaps it has not suffered the same tortures as mine.)

I felt suddenly free once these contraptions were removed. I had my private space back again. My room looked almost ordinary, like any other bedroom; no industrial strength extension cords lined the walls, no red lights winked at me as I crossed the room, no lamps turned on by themselves. Simple tasks, like switching on lights *after* I entered my room, or turning them off again *after* I was already settled in bed, were now enjoyable. The only alarm I needed to set was on my clock—and I resented even *that* mundane function!

Since I was psyched up, I decided to go one further: It was time to set some ground rules for overnight observers. Anna and I both needed our privacy, and we hadn't had much of it. I was, admittedly, feeling more confined and scrutinized, but then I'd previously enjoyed a more private lifestyle. Although I had become accustomed to this communal living arrangement, since moving in with the family, I still needed my own private space. My bedroom had become my sole refuge, but now that was being intruded upon on a regular basis and I was left with no retreat.

The aliens (or whatever they were) did not seem impressed by all the gadgetry; the abductions continued anyway. So why should I have to tolerate human intrusion as well?

35. Anna

Since the spring of 1992 we have been plagued by strange telephone service. We live in the country, where telephone service used to be rather poor, but it doesn't account for all these anomalies. We frequently have single rings on the line (I thought everyone got those), callers that refuse to identify themselves, calls where people refuse to talk, strange voices or humming on the line, and calls from purported government officials.

I first started noticing it in April of 1992. I found it bizarre at first, frightening when I thought about it later, and then became indignant. Now, I don't care anymore. My first inkling was when Beth's parents were visited by Air Force Intelligence in the middle of April and her father was questioned about Beth's involvement with "investigators of anomalous phenomena." I immediately called Richard Hall suggesting that someone's phone was tapped—mine, his or Rob's. They were the only people I had talked with, except of course the people we had met at Rob's apartment. Even though Rob's phone did the one-ring thing just after we arrived (he'd never had that happen before), Richard dismissed the people we met as having any ties to Air Force Intelligence. He suspected that a phone tap was possible. He had been convinced that his phone was tapped at various times over the years, but didn't have any indication that it was tapped now. Richard reassured me that we weren't doing anything illegal, immoral or seditious by exploring what was happening to us. He asked me to keep a log of strange telephone events since in the previous two weeks we had one phone call in an unintelligible language (tonal patterns interspersed with clicks), and a caller who only said, *"Don't,"* before hanging up. Don't what, we never figured out. So, dutifully, I started keeping a log.

During the next two months I recorded five instances of people calling on the business line (installed in late March and not listed in the phone book), asking for a nonexistent person, then immediately dialing the home line asking for the same person. Once the same caller asked for two different people. I answered one of those double calls and asked where the woman had obtained the numbers. She said she was a receptionist at a doctor's office and a patient had given her the numbers.

On July 9 and July 14 Beth received the two frightening double calls. We had become annoyed with them, so had begun making sure we knew what the nature of all calls were before we answered any questions. After all, this was a business line. One male caller only wanted to know who he was talking to. Beth didn't tell him any more than the farm name and asked whom he wished to speak to. He said he was with the FBI, but wouldn't state the purpose of the call. She hung up. Fifteen seconds later he was back on the home phone line. Same question, same response, same result. On the 14th, she had the same type of call, yet this time the man identified himself with the Secret Service, no stated reason for his call, just wanted to know to whom he was talking. When the house phone rang seconds after she had hung up on him, she answered with "This is still The Farm," he replied, "This is still the Secret Service." He would give no information, neither would Beth. Stalemate.

My question is, why are they harassing us? Whoever they are, I'm not convinced they were with these agencies as they claimed; anyone can make that assertion. They were frightening me. What purpose does it serve except to make us paranoid? Maybe that's the caller's purpose. Make people that say they have been abducted by aliens also tell of government interference, which of course would be denied, and thereby discredit everything they say as a product of a paranoid personality. I admit we were getting quite paranoid, but Beth's abductions were still happening. Why couldn't we be allowed to deal with one strange situation at a time? It was hard enough to remain functional. Maybe that was the point. But why?

In the middle of July Bob bought a Caller ID box for us. This way we would at least know the telephone number of the people who were calling and harassing us. Unfortunately, Caller ID only reveals local numbers (unless the long-distance caller also has a box), so that hope was dashed. Almost anywhere is a long distance call from the farm. We have a local calling radius of ten miles to the south and much less to the north. Bob hooked up the box and I called the phone company to have the service connected on the afternoon of July 17. The man said that there was a "special handling" code on the account and that he could not change the account. He then said that special handling was actually on the same phone number, but with a different area code within the same state. (On July 19 I called that number and got the message: "This is not a working number.") He then suggested that I use *69 (return call) instead of Caller ID. I explained that it did not work in our area, or so we'd been told a few days before. I was assured that it was a service available on our line, and that it should work. The operator would not explain what "special handling" meant except that one needed a password to access the account to make changes to the service. Only certain people in the main office could make those changes. Why did he suggest that I use *69 if "special handling" really was only on the other number that matched mine in another area code? Caller ID did not work on my phone.

On July 20 Beth talked with a telephone company supervisor about why Caller ID did not work. She said there were "special handling" codes on both the residence and business lines. They had been removed so that Caller ID could be put on. There was a code to have them reinstated after the service was changed. She assured Beth that Caller ID would soon be working. When Beth asked to talk with "the nice man that helped me on Friday," she was told "He is no longer with us." On each of the next two days, Beth called the telephone company to find out why Caller ID was not working. They kept saying they would look into it.

On July 24 I called. I was told that Caller ID was hooked up on July 21. Maintenance called back and said we probably had a defective box; it worked from his end. I called back and asked to talk with a supervisor. The supervisor said, "There are no 'special handling' codes on your lines. There's no indication that they had ever been put on. The man you spoke with must have made a mistake." She offered to write me a letter to that effect. I declined her offer. The next weekend a friend, who had Caller ID on his phone, took my box and tried it. It worked fine. Eventually Caller ID did work. So far, we have not had any more double calls or calls from people claiming to be government agents.

A good friend of mine worked at the phone company for over ten years, so I asked her what special handling codes were. She explained that these were put on telephone lines that were tapped, or in some cases when there had been a messy divorce and one partner wanted to keep the other from disconnecting their phone service. It was a service that only the account holder or a government agency, with court approval, could request. A special code had to be given to certain supervisors before the service could be changed. I certainly hadn't requested it. When I related my frustrations with the phone company, she explained that most people do not know about special handling codes, and they were taught very early in their careers not to mention it. If the calling party didn't have the code word, they were not to mention "special handling," and to deny that it existed. The man who told me about it may have been fired.

I talked to staff members in both my Congressman's and Senator's offices to see if they could stop the suspected phone tapping. Neither office felt it was something that they wanted to deal with. Come on. Illegal phone taps on one of their constituent's phones and they are not interested? I thought at least I could find out if the phones were actually tapped, and find out whether or not a court order had authorized it. It would have been interesting to see the reasons for tapping the phones.

Caller ID now works fine, although most of the harassing phone calls have stopped, so it doesn't meet one of my expectations: To find out who was doing the harassing. Our harassing calls are all long distance, of course (the box reads "out of area call"). Beth received another strange call the next week. Although she did not know it was strange at the time. A man called and asked for Bob Luca, and apologized for having dialed the wrong number. When she told me about it, I immediately connected that name with the name of Betty Andreasson's husband. I had read all of Ray Fowler's books on Betty's abduction experiences. Beth had read none. I looked in the local area phone books, but there was no Bob Luca listed in any of them. A week or so later I received a call asking for Oscar Jamerson. The only reason it was strange was because that was my grandfather's name. He died in 1964. It is not a common name. More harassment? I don't know.

In August Beth called our veterinarian on his cellular phone to come to the farm to treat a sick horse. She reached him, but had such a bad connection she couldn't understand him. The phone connection was broken. He called back in a couple minutes, but the connection was still bad and they were cut off. After being disconnected, Beth heard a male voice on the line say, "It's only her vet." When she tried to talk to the voice, she did not receive a reply.

What I don't understand about all of this is why these guys are so inept. If they wanted to monitor our phone calls, why did they make it so obvious? Our technology is advanced enough that I need never have suspected the phones were being monitored. Why did they bungle it so? They wanted us to know. Why? *Big Brother* is watching? My message to them is, "Why don't you come by and protect Beth from being abducted? Don't just sit on the sidelines and watch and listen. Help us!"

36. Beth

The Gathering. That's how we'd begun to see it. It seemed that abduction experiencers were rolling out of the woodwork, ordinary people no different than Anna or myself with remarkably similar stories to tell.

At first, I thought I must have been unconsciously influencing these people, dropping hints or making offhand remarks that they absorbed and then regurgitated in their own words. How else could I explain it? So many people of such diverse backgrounds couldn't possibly be congregating at the farm just so they could relate bizarre stories of UFO sightings and alien interference.

Not all of these people were strangers; some were boarding horses with us, some had attended clinics and seminars. These acquaintances could have been picking up something from me. At least that's how I preferred to explain it. But it was the others, the outsiders, who strained that theory. There were also visitors to the farm who had openly discussed eerie happenings long before I became aware of my own involvement or had any conscious memories of those earlier unexplained events. I could not possibly have influenced *these* people!

In the summer of 1991, a woman in her mid-thirties who had attended a four-day clinic at the farm, stayed for a campout with her nine year-old son. Not wanting to leave them to camp alone, I pitched my small tent alongside theirs, and we built a cozy fire a few yards away. Once her son was settled for the night, she and I sat before the fire and chatted quietly about nothing in particular. After a while, the conversation narrowed to a particular topic which seemed of interest to my guest: UFOs. Had I ever seen one? she asked. I nodded, but not wanting this woman to think I was a kook, I didn't offer to comment. Besides, my most recent memory of having seen a UFO was with Anna and her sister two years before—and even that memory was foggy, like looking through a mirror clouded by age. The subject was disturbing nonetheless, and I hoped she would not dwell on it too long.

Her experiences, she went on to say, were a little fuzzy; it had been years before. But lately, she added, odd things were happening to her and her son as well.

Despite my reluctance to discuss these things, I couldn't resist hearing more. "How does this relate to UFOs? I suppose strange things happen to everyone."

"I suppose so," she agreed. "But I really think it's all connected in some way. I don't know why I think that. You must think I'm crazy...."

"No!" I barked, much too quickly. I didn't think she was crazy, but I didn't think she was being rational either.

"My son has been having awful nightmares lately," she continued; "and me, too. We both dream about the same things; being floated up into some kind of craft through beams of light, seeing strange beings in our bedrooms. He even draws pictures of them, says they're not from *here*." She emphasized the word *here* by pointing to the ground.

I shivered, though it certainly was not cold that summer night before the campfire. My body felt numb and I was tempted to pinch myself to see if I was dreaming.

"*Have* you seen a UFO, maybe in dreams?" I ventured.

"Probably there, too. I remember seeing one—in real life—when I was about ten years old. I was sitting with my friend on a stone wall behind my house. It was summer and really hot, and we thought it'd be neat to go outside where it was cooler and just watch the stars." She paused briefly, as if trying to focus on the memory. "There was this really bright bluish light that we thought was an airplane flying over, but then it suddenly dropped way down and just hovered right over where we were sitting. I remember being very frightened, but not being able to move, to run away. Then a beam of light sort of floated down over us, like a solid thing rather than amorphous, and it was so bright I couldn't see through it."

She paused again, and my mind filled in the blank space with a dreamlike vision of a bright light hovering overhead—then it was gone.

"You know, I don't remember what happened after that! I guess we just went back inside and went to bed. Isn't that strange? I didn't realize the rest was such a blur."

She then told me of other, similar sightings and periods of missing time which she believed were connected; some as recently as the previous fall.

I hadn't spoken during her monologue, and once she had fallen silent the night seemed to have taken the cue and become absolutely still. The quiet made me jittery.

Less than a year later, a woman who boarded her horse with us came to me complaining of physical problems. She seemed to want a listening ear and asked if we might go out one evening and have dinner someplace nearby. During the meal, she described experiencing changes in her eyesight that had confounded more than one eye doctor. I was immediately reminded of one of my students, the young woman who had burned retinas from a blinding light, and who later claimed to have no memory of being injured. (She told me it was caused by an eye infection and had denied ever telling me anything else!) I didn't know what I was expected to say or what advice I could offer. So I just listened and tried not to make too much out of it.

A few weeks later, this same woman told Anna about "becoming lost" on her way home from the farm one night, though she couldn't recall ever having been at the farm after dark and driving home alone. She described her decision to take an unfamiliar route, not remembering anything she passed along the way, and arriving home much later than expected. For some reason, she admitted, the incident was still upsetting to her, and she had hoped one of us might be able to shine some light on it. (Excuse the pun.) We couldn't, of course, but the woman's story left us both feeling uneasy, as if we were waiting for the other shoe to drop.

Less than a month later, the other shoe dropped: A rather withdrawn woman in her late thirties had come to the farm looking for a well-trained, small horse for her ten year-old twin sons. I had expected her to be a difficult person to please, as she appeared so introspective and troubled. But I was determined to make a good impression. The best advertising is word of mouth, and even if the lady didn't buy a horse from us, she may one day recommend us to someone else who would buy.

We spent over an hour looking at the stock and finally found a mare she seemed interested in. Her sons were not with her, so she didn't wish to take up my time, asking if she might come back in a few days with her sons so they could try out the mare.

The following day she returned with twins in tow, both boys excited about the horse their mother had found for them. The boys headed straight for the stable, while their mother and I stood in the drive talking.

"I'm so relieved," she sighed. "They need something to take their minds off things."

I was curious about her remark, wondering if the boys, who I'd had little opportunity to greet, might be emotionally or physically challenged. I decided to just ask her, since she didn't seem the type to volunteer information.

"Oh, no," she responded, smiling. "Physically they're fine. It's just...well, Ken has been having some problems lately. Just bad dreams, you know?"

I nodded, but my body went into automatic yellow alert.

The mare proved to be just what they were looking for, the twins were typically enthusiastic and impatient to have her home with them. Leaving the boys to look around, their mother and I returned to her car to discuss details of the sale. Once that was done, she again brought up Ken's dreams, bravely adding that she, too, had been troubled by nightmares. I wasn't about to ask her to describe these dreams, and felt she was too shy to divulge anything further without encouragement.

I was wrong. She grasped my arm, startling me.

"I know you won't believe this, but I think aliens are in our home!"

I had been afraid of some associated revelation, but for this woman to come right out with something like that took me completely by surprise! What could have possessed her to tell a stranger such a bizarre story? Did she really expect me to believe it, to empathize or understand? Was I wearing a sign on my chest that read, *"Tell me your own personal UFO/alien story,"* or perhaps, *"I'm a sucker and I'll believe anything"*?

Despite this awkward beginning, we became friends and have since discovered many things about our pasts and our individual encounters with these alien beings. Ken has continued to draw what he remembers, as I have, and as his mother once did. The other twin has withdrawn from his involvement, expressing a profound fear of the "creatures who come into his room." There is another child in the family, a daughter of eleven, who asks pertinent questions about "ghosts" in the house and whether people can really fly, but otherwise seems more curious than frightened.

Our barn manager, a married woman with a grown daughter, has also admitted to unusual happenings. One morning, as we discussed the day's planned activities, she complained of being exhausted. When I asked her why she was so tired, she told of being kept awake the night before by a bright light shining through their bedroom window. The window, she explained, faced the barn (which was not wired for electricity) and the woods behind leading to the river. Light could not come from that direction, she had said adamantly. When she'd tried to wake her husband, he didn't respond, even when she nearly pushed him off the bed.

"He was like a dead man," she cried. "I know he can be a hard sleeper, but that was ridiculous! I even tried to open his eyes by pulling his lids up! Can you believe it?"

I was about to ask her if she ever found out where the light came from, when she rushed on:

"It's funny, but just when I was about to give up, I went back to sleep, like someone gassed me. Dropped right off. Probably wore myself out trying to wake the man up!"

I didn't ask questions, didn't ask for clarification. She indicated she thought it was a weird dream and expected to forget about it. She had told me, she said, because it was so odd. And didn't I think so?

Odd. Yes, it was odd.

A year ago I accepted two new students; brothers, aged seven and nine, who wished to take lessons together an evening each week. When they arrived with their parents the first day, I took them around the property so they could see where they'd be riding. As we walked around the cross-country course at the rear of the property, the mother pointed to the far corner of the field.

"What's that ring out there?" she asked me, shading her eyes from the bright sun, but making no move to investigate.

I stammered an answer, something about dead grass, hoping the woman would not want to go closer.

(Note: During investigations into our experiences, Anna and I have independently recalled seeing a craft in that particular part of the field on numerous occasions. When the area was later examined, the barren, raised ring was discovered. To this day, nothing will grow along that circular ridge.)

I deliberately picked up the pace, heading in another direction. As we passed the ring well to the north, I heard the boys laughing and turned to see where they were. They were running around the outside of the ring, laughing and waving at us to come join them. The parents immediately changed tack. I couldn't think of any way to stop them from going over there, so I tagged along, working hard to come up with an alternate explanation for the mysterious barren ridge.

"Wow!" the mother cried excitedly. "Looks like a UFO landing site!"

I stared open-mouthed at her. "A UFO landing site?" I repeated dumbly. "What makes you think that?"

"Oh boy," the father said under his breath. Turning to confront his wife, he berated, "Beth's going to think we're nuts!" He turned back to me, apologizing for his wife's outburst.

"That's okay." But it wasn't. More experiencers? A whole family, here? *Again?*

On future visits during the brothers' lessons, I learned through their own admissions that the whole family had conscious memories of being abducted by nonhuman beings and being taken aboard silent, hovering craft. As these intrusions continued, they began talking openly about their remembered experiences and investigating the phenomenon through books and other available material. They seemed like nice people within a normal, loving family—not the type to entertain fantasy. The parents were hard-working professionals who believed in a quality education for their boys that included a re-alistic view of their world. How did they manage to integrate this strangeness into their sensible belief structure?

What was going on here? Didn't people ever experience odd things that had earthly explanations? What ever happened to, "You'll never guess what happened! My car broke down again!"; or "I know you won't believe this, but we didn't get any rain!"; or "You're going to think I'm crazy, but I swear my pants shrunk!"

Has everyone moved onto the set of *Close Encounters?*

37. Anna

I was beginning to get paranoid. Everyone around us seemed to be experiencers. In July, I counted the number of local people I knew who I felt were somehow linked with these experiences. Friends and business acquaintances talked to us about disturbing dreams of strange creatures; had woken with unexplained burns, bruises or scars; reported vision changes; had been awakened by bright lights; had classic missing time episodes; or actually admitted to visitations and journeys with strange gray beings. Over 70% of the people (and usually their families) that did business with the farm seemed to be connected to the aliens somehow.

Were Beth and I attracting them? Was association with us caus-ing their experiences and fears? Were the gray shits picking on our friends? I didn't know what to think. I hoped not; I didn't see how it was possible. But it made me leery of advertising for new boarders or for riding students—aliens are bad for business. I didn't want to drag anyone else into this. Yes, I was definitely paranoid. Circum-stances would not abate my paranoia.

Beth had begun teaching riding lessons to a woman who had at-tended an exhibition at the farm in March. Joan felt that she had a confidence problem working with her recently gelded three year-old horse, and asked Beth to help her allow the horse to become tractable and easier to ride. Joan trailered her horse to the farm once a week from about thirty-five miles away. After their first lesson, Beth sug-gested to Joan that she was wasting her money; Joan didn't seem to have a confidence problem, and the horse was well-mannered. Joan decided to continue with the lessons.

During her third lesson, Joan asked Beth *the* question: "What's the strangest thing that ever happened to you?" This seems to be a common question from UFO buffs who want to seek out kindred spirits. I had read about this in several older publications. Beth skirted the question, since she had learned earlier that Joan's husband was an Air Force pilot. Besides, Beth didn't feel comfortable talking about any of this stuff with comparative strangers. When Beth told me about Joan's strange behavior, I suspected a spy! I was paranoid. I'd read too many books about the intrigue surrounding some experiencers.

A week later, Joan brought a "friend" with her who was supposedly interested in horses and had known Joan "for a short time." The friend (never introduced to Beth) had brought two cameras with her and took at least three rolls of film while Joan had her lesson. It seemed odd that many of the pictures were of Beth and the farm buildings, not of Joan and her horse.

The next week I was ready—I had my camera! I took pictures of her truck and trailer, her license plates, and then went into the arena and took pictures of Joan. She was visibly upset (and as a consequence, so was her horse), but I explained that I was taking pictures for a new farm brochure. This information did not seem to calm her. We talked afterwards and she volunteered the four pictures her friend had taken of her the previous week. She also explained to me that her husband worked for the airlines, not the Air Force, although I hadn't asked.

After her next lesson, Beth explained that she would be out of town for a week and would be unable to give her a lesson. Joan probed pretty hard to find out where Beth was going. She also called later in the week and tried to find out where Beth would be (at the Abduction Study Conference at M.I.T.). Beth finally did a quid pro quo—she lied, told Joan she was going on vacation to Maine! Joan then told Beth that she had to go to California for a couple weeks and would resume lessons when she got back. She never came back for another lesson. Was she a spy? I don't know, but in July we picked up a better candidate.

Sally attended Beth's week-long adult summer riding camp. Sally told us that she had been riding for quite a while and seemed comfortable around the horses. Unfortunately, it seems as if she only knew terminology and had few riding skills. Each time she got on a horse, she fell off (even at a walk). When Beth asked the students to do something, Sally watched the others to find out what was wanted.

It was almost as if she took a two-week crash course in terminology and riding and hadn't put it all together yet.

At lunch one day she also asked *the* question. She listened avidly when one student described a classic (to Beth) missing time experience, another recounted a dream of a deer being attacked by a lion that changed into two gray creatures, another recounted a strange car accident where there was no one around. Sally seemed less interested when another student told of a funny, embarrassing situation. She did not volunteer any information from her own experiences, nor did Beth. Sally did ask very personal, probing questions about other people's experiences, almost to the point of rudeness when the people didn't want to describe any more. She even trapped one student in the barn later in the day to get more information. Was she really an abductee looking for support, or a spy? It was time I tried to find out.

Sally was unwilling to talk about herself, but during the week Beth was able to gain some information. She said she was an Air Force brat, had spent most of her life in Europe, recently returned to this country, and worked for the Defense Department. I checked with the barn manager where Sally said she'd been taking lessons, and they didn't remember her. Sally said she rode weekly with a horse club I'd helped to found twelve years earlier—they don't have weekly riding lessons and I'd never seen her before. No other club member I asked had ever seen her at activities, but her name did show up on a recent membership list. She said she rode dressage at the novice level. But that terminology (novice level) applies to a different riding discipline. The barn where she said she took dressage lessons teaches only hunt seat riding.

Sally's car intrigued me. The windows were covered in dark material and the car was alarmed! She was either very paranoid, or had something to hide. Abductee or spy? Each morning she parked the car, set the alarms, and then checked that movement within a foot of the car would set off the alarms. Our dog set them off one day. Intrigue continued when, the second day, she removed all the stuff hanging from her rear-view mirror, including her ID card. She had a "Blackhawk" sticker on the bumper that may indicate Air Force Intelligence. I called her work number and got a man that didn't identify the company. I called back later and got an identification of B.D.M. (a defense contractor implicated in some UFO-related business). I took pictures of her license plates, stickers and car (being careful not to set off the alarms), but never did anything with the information.

During the riding camp that summer, several students approached Beth and related incidents of Sally cornering them and asking probing questions about the farm and any strange happenings. One student admitted to thinking she was a spy of some sort. Sally decided not to attend the end-of-camp dinner when she heard that I was also going. The only question she asked was whether the farm would be unattended that evening before saying she couldn't make it. Beth told her yes, even though my sister and brother would be here. No one visited the farm that evening. We've not heard from Sally since.

We still occasionally have black helicopters (without identifying markings) fly over the farm and hover over buildings, are followed to and from the farm by men in dark colored cars, and have white panel trucks or camouflage-colored vehicles with radar discs (radio antennae?) atop them drive up and down the road in front of the farm. Much of this activity seems to be correlated to an abduction the night before. We sometimes joke that they know more about our abductions than we do. When we see the trucks in front of the farm, we try and figure out if we missed something weird happening the night before. Sometimes we have.

Again, why? What knowledge do they hope to gain by sending people and machines to spy on us? I don't think we've told them much. And why are they so clumsy and obvious about it? A friend of mine came up with a plausible explanation: Even if the government is interested in what's going on, it may not be their highest priority. They don't commit their best people and most sophisticated equipment to the project.

So, am I paranoid? You bet, and with good reason. What can I do about it? About as much as I can do about the gray shits! Nothing.

Joan did drop by about six months after her last lesson to let us know that she and her horse were doing well together. As she was leaving, she asked Beth, "Are the little gray guys still bothering you?" Beth told her she had no idea what she was talking about.

38. Beth

Four whole days! It was almost too good to be true. Two years had passed since my last vacation, and that was spent with my son and his family, leaving no time at all for me to relax in privacy and just vegetate! I desperately needed time alone, away from the farm and my students, away from investigators and other abductees, away from *people!*

And it would be a different sort of vacation, too. I had reserved a log cabin at a state park only a couple hours drive from the farm. Its closeness to home, however, did not discourage me from going. The cabin was situated on the side of a mountain ridge; it was remote enough to guarantee privacy and peace, yet close enough for Anna to come visit if she wished to get away for a bit, too. Since I had scheduled my vacation for September after the Labor Day crowds were long gone I could be assured of solitude with no one else around. Just the thought of it made my mouth water! I could hardly wait to leave.

Anna and I had been under tremendous stress with this abduction business. It seemed that every other night something was happening. Our sleep patterns were so disrupted it was hard to function during the day for more than a few hours. I had gotten into the habit of taking naps during the day, though before all this activity I had found it difficult to so much as close my eyes while the sun was still up. Anna was in as bad a shape, but for her, sleeping during the day was no problem. She had trouble just sleeping through the night.

A vacation might allow me that much needed rest and diversion. It wouldn't do Anna any harm either, if she could manage to take a day or two off work and leave the farm in capable hands.

I left home on Saturday of Labor Day weekend, 1992, my car packed with every conceivable necessity. Since check-in time was not until 4:00 P.M. Monday, there was really no need to leave on Saturday, but I thought it would be a good opportunity to spend the weekend with my parents, who lived only a half hour from the park.

It took much too long to get to my parents! I couldn't account for over an hour of time, but had no memory of anything having happened that would have delayed me. When I arrived at my parents' home, I felt tired and out of sorts, but believed it was due more to accumulated anxieties than anything else.

When Monday morning came, I was eager to get on my way, even though it would mean I'd arrive too early to go directly to the cabin. My parents decided to follow me in their car, suggesting we spend the day together roaming the park until it was time for me to check in. After an enjoyable lunch at the lodge, I checked in and the three of us drove on up to the cabin. My parents helped me unload the car, then left soon after. My father would have liked to stay awhile, but my mother was uneasy about the cabin and its location. I couldn't understand her reaction; the cabin was small, true, but it was cozy and clean, and no more remote than many areas where we had camped throughout my life. I had thought the rustic little cabin would bring back good memories and that they would want to stay as long as possible.

Our family had been camping since I was only two weeks old. I can recall, as a youngster, spending nearly every weekend (when my father wasn't working) camping outdoors with nothing but our sleeping bags, a pup tent, and enough food for two meals a day, which had to be cooked over an open fire. Later, my parents organized a camping club, which eventually totaled more than a hundred people, singles and families alike. Even when I was entering my teens, I still looked forward to those weekends.

It wasn't quite as rustic; the campsites encompassed many acres and included large mess halls (usually seating two hundred or more), single and family cabins, bath houses, lodges and lakes. There were miles of trails through beautiful woodlands, boats to row out on the lakes, and square dancing on Saturday nights. As I grew up I naturally developed other interests, but I was a dyed-in-the-wool, completely fanatical *camper*, too. It was a love of the outdoors and nature itself, and my parents' enthusiasm, which *kept* me interested through the years.

So I was surprised when my father so quickly conceded, ushering my distraught mother to their car. They drove off without a backward glance and I was left alone on the mountain—happily, gratefully alone. I dashed back up to the cabin, plopping down on the porch rocker, hardly believing I was actually there. Four whole days! Funny how vacations just flew by, as if those days were specifically designed with fewer hours in them. Time was already slipping away—precious irretrievable vacation time—while I sat musing on the porch. I pulled myself up and went inside to put away the groceries.

I ate a light dinner, then sat on the porch inhaling the pungent woodsy smell as an evening shower pattered the leaves and dripped lazily off the roof eaves. I wasn't disappointed by the unexpected rain; it cleaned the air after a warm day and calmed all manner of creatures, including this human creature. It was like being the last person on earth, and rather than being unnerved by that symbolism, I was rejuvenated. It was time to take that long awaited deep breath and allow the body and mind to heal.

That first night passed uneventfully and I awoke feeling refreshed and contented. I lounged in bed much longer than I would have normally, listening to the leaves rustle softly in the morning breeze and watching the reflection of the rising sun as it dripped across the window sill like liquid gold. I was in no hurry to get up, wasn't yet hungry for breakfast, and I hesitated to plan activities that might take me away from the cabin for an extended period; Anna had planned to be out early that afternoon after taking care of some things at the farm in the morning. It had taken some doing to convince her it would do her good to get away even for just a day. But I was sure she wouldn't be sorry. She was going to love this place!

I sat up and stretched, kicking off the heavy blanket which had become too warm. Hearing a loud buzzing outside the window, I turned to see what kind of insect was practicing its morning maneuvers. A gigantic furry bee dived-bombed the screen then ricocheted off and flew away, appearing none the worse for wear. Fully awake now, I got up and prepared breakfast, taking it outside to eat on the porch.

I wasn't very hungry; after only a few bites I put the leftovers on the floor. A chipmunk peering shyly around a tree stump seemed interested in the cornucopia I left behind. Scrambling up to the porch stoop and sniffing the air for danger, the little animal seemed to be planning a raid, perhaps thinking it might sneak off with the goodies before I noticed it was there. I nudged the plate with my foot, moving it a few inches closer to the hungry little beggar.

The chipmunk, apparently unable to restrain itself any longer, leaped onto the porch and bravely approached the offered meal. It nibbled delicately for a time, ignoring my presence. Suddenly the animal snapped alert, as if realizing at last that it was not alone, spun around in mid-air and disappeared off the porch. I hadn't moved at all and wondered what had startled the thing. Then I heard the buzzing again, louder than before, and saw the familiar huge insect circling the porch as if guarding its territory from would-be trespassers.

I couldn't believe the size of it! It looked more like a fat bird wearing an oversized fur coat than a bee. But I was sure that's what it was. It was uncommonly round without the usual divisions of thorax and head; it had no markings or stripes but wore the most brilliant golden downy fur! Admiring its color while respecting its size, I carefully rose from the rocker and followed it. It flew around the back of the cabin then stopped in mid-air and hovered like a helicopter awaiting permission to land. Shading my eyes from the sun, I studied the beast as it floated barely a foot away.

"Okay, Goldie," I dubbed it. "Have I taken up residence on your home turf?"

The bee fluttered as if revving its engine and I felt the hair on the back of my neck stand up. What kind of bee was this? There was something eerie about it; it was almost as if the thing actually knew what I was saying. Of course, that was ridiculous. It was only a large bee after all.

I retreated to the porch, keeping watch over my shoulder in case Goldie decided to charge from the rear. The chipmunk had obviously taken advantage of my absence to steal a few choice morsels from the plate, so I left the dish sitting on the porch floor and went back inside to refill my coffee mug. When I returned, Goldie was hovering just outside the screen door. I was beginning to get nervous, though the insect had made no threatening moves and quickly backed up to allow me to pass. I watched it from the side of my eye, moving cautiously toward the rocker. As soon as I was seated, Goldie flew directly before me, stopping at eye level as if wishing to study me at closer range. Ever so slowly, I raised my hand and waved it in front of my face, hoping to frighten it off. Goldie was not impressed, though, and simply moved to one side.

This posturing went on for some time. Periodically Goldie would fly off only to return a few minutes later, but even though I couldn't see it, I could always hear its thunderous buzzing. When I decided to take a short hike in the woods around the cabin, Goldie accompanied me; when I returned to the cabin for lunch and to wait for Anna's arrival, Goldie lingered just outside, zooming from window to door to window, depending on where I was in the cabin. When I was in the bathroom where Goldie was unable to keep me in view, I could hear it buzzing frantically around the perimeter, going from one window to the next. This was unnerving! How had I managed to adopt this beast? Most animal lovers were enamored with stray dogs and cats; I seemed to have attracted a love-sick bee!

The afternoon wore on, but there was no sign of Anna. She could have run into problems at the farm and was delayed leaving, I suspected. There was no chance I'd be lonely, anyway, with Goldie glued to me like an ornament on a Christmas tree.

Sitting on the porch, a cup of coffee resting on my knee, Goldie and I watched black clouds building over the mountain ridge. The moist odor of an impending storm lay heavily on the air; the hissing of wind-stirred leaves announced the coming rain in no uncertain terms. Then the rain came down in torrents, obliterating the view from the cabin's porch until only the closest trees were visible, lashing the branches of their neighbors. What a storm! I doubted that Goldie would hang around while the storm was raging, and I was right. The bee was nowhere to be seen.

The storm gradually wore itself out. By sunset the rain had stopped, leaving a heavy mist in its place. Goldie had not returned, and I was surprised to find that I missed the fat pest. I stood up and stretched, entertaining thoughts of dinner, when I spotted an odd fog bank on the ridge above the cabin. It was stationary and opaque as if painted onto the scenery. It was about six yards across and hugging the ground; there were no other tendrils of ground fog that one might expect to see under similar conditions. I felt uneasy watching it.

The rest of that evening was a blank. I did not recall eating dinner or going to bed. My next conscious memory was waking up the next morning feeling very tired. My stomach was upset and I feared I was coming down with something; I'd probably gotten a chill from sitting outside during the storm. Fate wouldn't be so cruel as to inflict me with a cold on my vacation! I threw off the covers and realized that I was still dressed in jeans and sweatshirt, but at least had had the wherewithal to remove my shoes before getting in bed.

Determined to make the best of it, I stumbled to the bathroom and turned on the cold water tap. But there was no water, only a few drops that plopped into the enamel sink like miniature transparent slugs. I went into the kitchen area and tried that faucet, but it was dry, too. Why was the water off? Ignoring my grumbling stomach, I hurried to change into fresh clothes, intending to drive the two and a half miles down to the lodge to notify the park custodians of the problem.

When I removed my jeans I understood why I had felt only half dressed earlier: I was not wearing any underpants. Why would I have gone to the trouble of taking them off then putting my jeans back on before going to bed? I must have been feeling worse than I thought! Shaking off an ominous sense of having been violated, I quickly dressed again and grabbed for the car keys which I'd hung on a nail behind the door. But they weren't there. I must have left them in the car, I thought, stepping out onto the porch.

It was a glorious morning. The sun was shining brightly, reflecting off the wet leaves that littered the walkway leading from the cabin to the small graveled space where my car was parked ten feet out into the middle of the roadway! Dear God, I moaned. How had that happened? Did it pop out of gear during the night and roll backwards? I seldom set the emergency brake, so it could have happened. Fortunately there were no other people on the mountain, so thankfully the car hadn't blocked anyone's passage.

I brushed off clumps of wet leaves from the sides of the windshield not bothering to clean the smudged glass in the center; the wipers would take care of that. This done, I reached for the door handle and froze. A sequence of startling events projected itself through my mind's eye. Suddenly I was no longer standing beside my car, my fingers gripping the door handle. I was no longer concerned with not having running water, no longer wondering why Anna hadn't come. I was someplace else entirely.

I saw the fog bank begin to move. It didn't drift like fog is expected to drift; it didn't break up or show signs of dissipating. Instead it remained compact and simply "walked" itself down the slope, moving (it seemed) in a deliberate fashion, avoiding trees and flowing over fallen limbs and downed trunks like whipped cream over a warm slice of pie. I watched it curiously, ignoring the tingles that ran up and down my spine—a warning that something was wrong with this scene.

Suddenly three figures stepped out of the fog bank and continued on down the slope toward the cabin. The fog immediately vanished. I thought these small creatures, gray and hairless, looked frighteningly familiar. I knew that whatever happened I would not remember it anyway, yet I didn't then understand why I was so confident of it. The approaching beings triggered my flight response and I jumped up off the rocker and fled into the cabin. Subconsciously aware that this evasive action would not save me from whatever these creatures represented, I grabbed my keys from behind the door and made a dash for the car.

Fortunately, the car was not locked so I jumped in, making a conscious effort not to look up the hill to see how close they were. The windshield was covered with leaves and I knew I wouldn't get far until I had cleared a space large enough to see through. I quickly started the car and switched on the wipers. The blades grudgingly pushed the leaves aside, and through the smudged glass I saw the three gray beings in front of the vehicle. One moved forward and raised his arm, pointing an object he held in his hand at the car's hood. I heard a popping sound, but didn't wait to find out where it came from. Jerking the gear lever into reverse, I stomped on the gas, but the car only drifted backward. I stepped on the brake, fumbling for the gearshift, and was startled to find it completely loose. It felt like I was holding onto a broken tree branch. I didn't know what to do! The three grays were still in the same position as if confident I was going nowhere.

. Then I heard a distinct ping and a thought entered my head which I knew couldn't have been mine: *Come with us.* The thought was also disconcertingly familiar, and I knew they would do whatever was necessary to keep me from getting away. As if in a dream, I turned off the ignition and got out.

"Don't do anything else to my car!" I ordered. "I'll go with you, but I want to remember this one," I pleaded. "I have a right to remember!" There was no response, no indication from them that they had understood what I'd said.

I began moving back toward the path. The grey who appeared to be in charge floated out in front, the others took up positions on either side of me. We made our way back up to the ridge where I had first seen the fog bank, then came to an abrupt halt. Instantly we were engulfed in a blinding blue-white light. I couldn't see through it, couldn't feel it, yet it seemed like a solid thing. Unable to move my head more than a fraction of an inch, I strained to look up, hoping I could see what had produced this light, but there seemed to be no end to it.

As quickly as it had appeared, the light winked out, leaving me suspended over a dark hole. Then the darkness beneath me became part of the floor and I was permitted to stand on my own two feet. I hadn't had any sensation of movement, so didn't know whether I was still somewhere near the cabins, on some kind of craft, or in another state or country! I wasn't sure how much time might have passed; it might have been only seconds or it could have been hours. I should have been afraid, but was more interested in the place than in how I might have arrived there. It almost looked familiar. Had I been there before? I looked around cautiously and saw that I was in a large, brightly lit room, one of the small gray beings hovering a few inches off the floor to my right.

Another gray-colored being approached and stood before me, studying me. This one was much taller, over five feet, whereas the "escorts" (who had accompanied me to this place) were barely four feet tall. (See Figure 3.) The grey was so willowy I wondered how it could stand upright. It's head was oversized, like that of an infant, and seemed perched precariously on a long, pencil-thin neck. The being had huge black almond-shaped eyes, rather wide-set, that didn't reflect the room's light. It was as if the creature's true eyes were covered by snug-fitting black velvet. This being looked familiar to me, yet I couldn't remember where I might have seen him before. Him? How did I know it was male? There was no outward indication of sex, no visible bulges or mannerisms that would identify it as distinctly male or female, yet I perceived it as male and as an authority figure. The recognition was so strong that it nearly overwhelmed me. Why was everything so familiar? How could I have ever seen anything like this before?

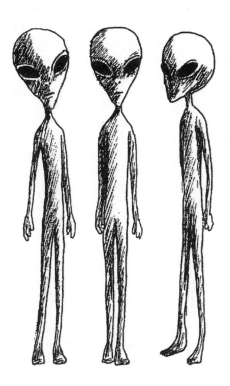

Figure 3. The Escort Service.

He "told" me to follow him, so I did as asked, beginning to feel ill at ease. Why was I doing this? We proceeded, by floating rather than walking, to another large room where I was told to sit in a molded chair that looked something like a giant marshmallow which had been molded into the shape of a recliner. The chair was cold to the touch, but not uncomfortable. As soon as I was seated, it conformed to my body from neck to ankle. This action seemed designed more for comfort than restraint as it didn't squeeze or grip, but merely adapted to my body's shape. There seemed little need for restraint anyway since I couldn't move anything but my eyes.

I detected motion to my right and saw three more of these taller beings, all identical in appearance to the one who had led me into the room. But these others I didn't recognize, and that confused me.

They all looked as if they had been poured from the same mold, like dolls lined up on a shelf in a toy store. It didn't make any sense that I'd recognize one individual over the others! Were these others male, too? I didn't know, couldn't tell. Not only was the sex indistinguishable, but also the structure. They didn't appear to have any bones or joints, no muscle definition whatsoever. (They moved by floating an inch or two above the floor, so perhaps muscles and joints were unnecessary. If bone mass was minimal or even nonexistent, it fell to reason that their environment was likely quite different from ours.)

The grey who led me in appeared to be in charge, as the others remained at a distance until signaled (in some unseen manner) to approach. The same grey instructed me to "not be afraid," that I would not be hurt. I heard these words in my head and knew he had not spoken them aloud. He had not moved his mouth, which remained a closed slit, yet I heard him clearly and knew I had heard these same words the same way many times before. Then he leaned down close to me and stared directly into my eyes.

I tried to shut my eyes against this invasion of my space, but couldn't. I tried to scream, but couldn't even open my mouth. It was as if my entire body, inside and out, had turned to stone. I worried that I wouldn't be able to breathe! Then I understood. This was the Doc. (See Figure 4.) I knew him because I had given him that name long ago. He straightened once again and moved out of sight, leaving me to wonder how long I had been associated with these beings, and what type of relationship had been established. What had they done to me over the years, to me and my family? It was all beginning to fall into place, memories sorting themselves into recognizable events…. But if I had been made to forget before, would I be permitted to remember this?

I felt my right arm being rearranged in the soft folds of the chair so that the palm of my hand was turned toward my body. I glanced to the right and saw one of the other greys holding what looked like a dual-chambered syringe with two needles shaped like a prong. There was a tube extending down from the end of the syringe. Both chambers were filled with an amber-colored liquid, one slightly darker in color than the other. Before I had time to register the obvious significance of this, the grey injected both chambers into the fleshy area of my hand just below the thumb. There was very little pain because the entry holes were already there! (After a period of missing time during the drive out to my parents' home the weekend before, my mother discovered two puncture wounds in my hand in this same area.)

Doc stood
between 5'3"-5'5"

Figure 4. Doc, the grey present during most of Beth's abductions.

I complained bitterly of the pain, even though my protests were non-verbal. The response came immediately and in a form I at once understood: They were using some form of telepathy! "There is no pain," I heard him say. I heartily disagreed, but heard again, "You have no pain." When the prong was removed, I mentally cried out and felt a hand press down on my forehead. The voice filled my head: "There is no pain." And there wasn't! In fact, I felt nothing at all. I was completely numb, almost giddy, as if I had drunk just enough alcohol to get happy, but not enough to feel disoriented. Two more injections, preceded by a popping sound, were "fired" into my upper left arm, but I felt nothing. I didn't care at that point what they did to me!

The syringes were withdrawn and my clothes removed while I remained in the chair. Still on my back, the cold from the chair more noticeable now, one of the attending greys, which I began referring to privately as technicians, or "techs," leaned over my abdomen and appeared to be examining my navel. Still feeling somewhat detached from these proceedings, I wasn't yet worried about what they might be planning to do down there. On a deeper level, I think I already knew what was coming. A sharp needle-like instrument was produced and I felt a hot pain as it punctured the skin just above my navel. The pain quickly subsided, replaced by a feeling of pressure, as if I were being blown up like a balloon. Eventually this instrument was also withdrawn, only to be replaced by yet *another* needle!

Mentally I cringed, my mind spewing out demands that they tell me why they were doing these things! I heard many answers, but learned nothing: "It is not important for you to know. You do not need to understand. It is necessary to do this." *Do what?* I insisted. What exactly were they doing? "It is part of the change," I was finally told. What change? This received no answer at all.

Now my lower left arm was being examined by one of the techs. The syringe was positioned just over the vein inside the elbow, the needle injected, and blood drawn slowly into the syringe chamber. Suddenly there was a stinging pain and the needle was quickly removed. The tech stood upright, still clutching the syringe. I saw Doc move toward him and stop. He looked directly at the tech and the tech disappeared! He instantly reappeared across the room to line up beside the other techs, who apparently had not been assigned a task and were patiently awaiting orders. What happened? I asked Doc. "The vein has been collapsed," he answered, then immediately summoned another tech to complete the job. This time, my other arm was used and blood was withdrawn without mishap.

I was left alone for a time and may even have drifted off to sleep. When I opened my eyes, Doc was back. He remained off to one side, but told me to look at the screen in front of me. I saw a twelve-inch square of clear material, maybe a half-inch thick, suspended before me. On it (or perhaps inside it) were a dozen or so horses moving about at liberty. There was no background scenery, no sound. I couldn't determine where the image was coming from, whether it was projected or contained within the screen.

Doc told me to watch the horses and understand that they had been changed. How? I asked. What did he mean, *changed?* There seemed to be no more information coming, so I watched the horses, not understanding what it was I was supposed to glean from this display. Within moments this screen was replaced with another, this one showing cows. The cows, however, *had* been noticeably changed. There were tubes protruding from the stomachs of all the cows! The tubes reminded me of a clothes dryer's exhaust hose, except that each one was capped. The cows had also been changed, Doc announced. That seemed obvious, but my questions on why went unanswered.

This screen was removed and the Doc once again leaned over me and stared into my eyes. I "saw" images of both horses and cows, then nothing. Immediately I heard, "You must only eat cow things." I was sure I had misunderstood! But then it was repeated several times. Finally he pulled back, but I had to understand what that meant. "Cow things?" I asked him. What did that mean? Didn't they understand that humans couldn't subsist on just beef and dairy products?

This seemed to hold little concern for him, as he only repeated the command that I "eat only cow things." Then he added, as if it were an afterthought (or perhaps because I was thinking the same thing), "You will not be crazy." That was certainly up for debate!

Abruptly, I was informed I could go, that they were finished with me. My jeans were pulled on with obvious difficulty minus my underpants. Next came my t-shirt which caused them almost as much trouble. I was impatient to go and frustrated by their clumsy efforts at dressing me. I demanded to be allowed to finish dressing myself and was surprised when they agreed. The greys retreated and the chair began to move. I felt myself being lifted upward, then unceremoniously deposited in a standing position. The chair released its hold and returned to its former state.

My sweatshirt lay crumpled on the floor beside the chair, but it might as well have been in China. I was still immobilized and couldn't pick it up! I asked to be released from this state, but received no encouragement. There was a brief pause while the greys seemed to be mulling it over, then Doc made a palms-down motion with his hand and I found I had limited mobility, enough at least to put on my sweatshirt and shoes. Evidently they had decided I was an acceptable risk as long as I was not given complete freedom.

Once dressed, I braved a step toward the greys, testing the boundaries. Instantaneously they raised their hands as if warding off an evil and unpredictable threat. I was again paralyzed.

For just one glorious moment I experienced an unexpected rush of power! They had been able to control me, yes, but if I were permitted complete freedom of movement and decided to take advantage of it, I could easily overpower *them*. They could be hurt, and somehow that knowledge made me feel a little less victimized.

The memory flashback, filled with bizarre images, held me hostage for over an hour. When it ended, I stumbled, nearly falling over. It felt as if I had been standing stock-still for days! All my muscles seemed suddenly to have liquified, and I had great difficulty forcing them to respond. When I was finally able to move, I crawled in behind the wheel of my car and attempted to start it, but nothing happened. I couldn't get it in either neutral or park, and it wasn't going to turn over while the gear was in limbo. Being unable to start the car, I was forced to abandon it.

I had no idea what was going on, why I had those strange images and memories. My thoughts were a mass of jumbled wires short circuiting all through my mind. I knew I was about to panic, but somehow managed to pull myself together. First, I had to get down to the lodge. I needed the reassurance of normal people, and I needed to locate a mechanic to repair my car. I certainly couldn't just wait for someone to come to me! As for those crazy images, I didn't have the courage to deal with them right then.

Steeling myself for the long walk ahead of me, I started down the mountain using the trails I had explored earlier under much more pleasant circumstances. After only a few minutes into the hike, I heard a familiar buzzing sound: It was Goldie, flitting around my head, dipping and soaring in true acrobatic splendor. Even that strangeness ceased to affect me anymore. I trudged down the mountain with Goldie at my elbow most of the way, although my companion made itself scarce as soon as I was within sight of the lodge.

I made the necessary call to a garage near the park, then advised the office of the water problem. I was told that after a heavy rain the runoff and leaves often clogged the system, which was admittedly dated, shutting off the water to the cabins, but I was assured maintenance would take care of it shortly. I started back to the cabin, this time taking the road in hopes of a ride by either the maintenance crew or the garage mechanic. Goldie, as expected, was waiting, staying with me all the way back to the cabin.

The tow truck arrived within the hour, though the mechanic was not encouraging. The car would have to be taken to the garage for repairs since it appeared the transmission cable had snapped and some of the electrical wiring was damaged. He asked me if the engine had caught fire, showing me where the fire wall was scorched black. I shook my head, not having a clue how or when that might have happened.

Exhausted from the morning's events, and the long hike back and forth from the lodge, I slept away the rest of the day, waking only periodically with cramps and nausea which I assumed were a response to the stresses of the day. Anna did not arrive that day either.

Just after sunrise I made another trip down to the lodge to call the garage and find out what the verdict was on my car. As usual, Goldie accompanied me. The mechanic assured me the problem could be fixed, quoted a price (which was much more than I had expected), and promised it would be ready in a week or so.

"A week or so?" I asked, incredulous. "I'm on vacation and I have to check out Friday morning before ten! How am I supposed to get home?"

The mechanic coughed into the phone. "Well, ma'am," he drawled, "I can't fix it overnight. Can't you call someone to come get you? You said you live a couple hours away, right?"

If I lived in France it wouldn't have made things more difficult! I was on vacation and it was turning out to be the most frustrating, confusing, annoying, *expensive* four days of my life! "Tell me," I demanded, "why should it take so long to replace a transmission cable and a few wires?"

He explained that they had to order the cable since it was a foreign car and they didn't have parts in stock. "And, uh," he stammered, "we have to replace the burned wiring first. Nothin' on this car works, ma'am. Now you can take it somewhere else if you want, but they'd have to do the same things if you ever want to drive it again."

Not having any alternative, I told him to go ahead and make the repairs. I would call Anna and have her pick me up, then come back the following week to retrieve my car.

Anna arrived Thursday evening just before dark, explaining that she had been tied up and couldn't leave sooner. She had also had car trouble and had to borrow her sister's small pickup. If I hadn't called, she admitted, she would probably not have come at all. As it was, we would have to pack everything in the bed of the little truck and pray that it didn't rain. I filled her in on the car situation, told her about my flashback (or whatever it was), and described my persistent companion, Goldie. Although Anna never got a good look at Goldie during her brief stay, she did hear the insect's loud buzzing and caught a fleeting glimpse of it outside the cabin's screen door. It appeared that Goldie's curiosity did not extend to actually meeting my friend face-to-face, and it was nowhere to be seen when we packed up and left the mountain.

The following week my parents gave me a lift to the repair shop so that I could retrieve my car. When I arrived, I was informed that much of the interior lighting still didn't work because the wiring was fused to the metal!

The mechanic looked at me askance, mumbling under his breath, "Must have driven it through a blast furnace...."

I felt like I should have told him *something,* offered some sort of plausible explanation for how these damages occurred, but then thought better of it. What could I have told him? That a bunch of disgruntled short, gray alien creatures zapped my car with some kind of ray gun? Or maybe that while the car sat parked and unused for two days, it got bored and did it to itself for amusement? Instead, I just shrugged.

I drove the car back home without incident, but worried over every little creak and click it made. I decided that if something similar should happen again, I would remember to steer clear of my car! I didn't think the greys would kill me, but evidently they had no qualms about killing cars.

For several months following this episode, I suffered with the "eat only cow things" diet. Initially I couldn't tolerate any products that were not derived from beef or dairy cows. I told myself that this was an implanted suggestion only and that if I refused to accept it as truth I could overcome it. Eventually I was able to eat vegetables and fruit, but it was some time later before I could handle other meats without becoming extremely ill.

(Note: Early in 1994 I learned that another abductee, who lived in the southwest, had an experience very similar to mine. While camping overnight with friends, she had been pestered by a flying insect that neither behaved nor looked like any insect she had ever seen. Evidently this insect followed her everywhere much like Goldie followed me but promptly disappeared right before the abduction took place.)

39. Anna

About two weeks after Beth's full conscious memory of an abduction, we learned that Budd Hopkins would be speaking at a local conference. Richard Hall arranged for us to meet with Budd, and Budd agreed to hypnotize Beth to explore that same incident.

Basically, the same information came out under hypnosis as Beth had consciously recalled, although there were a few more details. We do not have a tape recording of this session because the tapes that were used on Richard's recorder (Budd's malfunctioned!) have since disappeared. Afterwards, when we were discussing the incident, a light bulb exploded right next to Beth and I.

I always knew that Beth was being abducted, now maybe she'd believe it, too. No such luck. After her hypnosis session with Budd she seemed to need to deny it even more. Me, I was jealous.

I so wanted to find out about my own memories and fears, yet there was never enough time for me. Beth's needs took precedence over mine. It wasn't okay, but I needed her functioning enough to take care of the farm while I attempted to earn my pay at work to continue to pay the bills. Anyway, Budd had promised to continue to work with us, so I could wait a bit longer.

I did get to talk with Budd about my continual headaches as I drove him back to town to meet with another abductee. He suggested that the headaches may be caused by repressed memories. He had met other abductees who had similar problems and their headaches eased as they got more memories. I could hope. He also suggested that I give up reading about UFOs and abductions. He felt that I may be trying to force my memories, thereby contributing to the headaches. I reluctantly agreed to stop reading.

Budd also counseled me to find out if there was any organic cause for the headaches. Even if I thought my physical problems were caused by alien interactions, I needed to be sure that there wasn't a more routine explanation, one that could be cured by terrestrial medicine. I needed to stay physically healthy to deal with the mental stresses.

chapter 8

Mutual Pasts

40. Beth

I hadn't been on a train in many years, not since my early twenties, so I was anticipating this trip to New York like a kid about to go on her first train ride. The weather was cool, but not as cold as expected for that November morning as we boarded AMTRAK out of Washington's Union Station. The sun was shining and we were eager to get started, yet both of us were absorbed in our own thoughts.

I looked forward to seeing Budd Hopkins again and hoped the impending sessions scheduled for the next three days would clear up some of the confusion over recent events, perhaps even explain them. Although the September hypnosis session had apparently verified the cabin experience as having actually occurred as I had consciously remembered, it might still be a product of my imagination, those imaginings merely being reinforced by the hypnosis process. I knew about confabulation, how easily the mind could deceive itself, and realized that this may still be the case. Knowing this and being prepared for it made me all the more determined to get to the truth of the matter, even if it turned out I'd made it all up.

A mingling of tension, anticipation, fear and confusion thrust me into a state of physical distress as the train carried us closer to New York. I experienced stomach upset, diarrhea and dizziness, even though I kept reminding myself that I had no reason to feel so disturbed. As lunch time approached and we made our way to the dining car, I was sure I wouldn't be able to eat a thing. I did, however, manage to force down a rubbery hot dog from the snack bar and something to drink while Anna suffered along with me. Returning to our car an hour later, I actually felt better, beginning to understand my qualms. I wasn't really concerned about what we might discover under hypnosis, I was worried about what we might not discover.

Up to this point, I hadn't spent much time considering the other side of the coin. I had been looking for logical and rational explanations for these experiences. But what if there weren't any? What if we found no explanations and had to start dealing with these things on their own merit? *What if they had merit?* Relating my fears to Anna, I was surprised that she had similar thoughts, but had been afraid to mention them to me, feeling I was too vulnerable at that point. But these things had to be discussed openly, had to be considered possible, if we were going to prepare ourselves for the outcome of this self-examination.

Hypnotic regression used to examine abduction memories had been under heavy attack for some time, long before Anna and I became aware of this phenomenon. I was not afraid of hypnosis. Having been hypnotized years before, I knew there was no danger of my being "left" under or compelled to perform some unnatural act while in a trance state. Nevertheless, the process could cause damage, though admittedly that damage was more in the form of leading by the hypnotist that could produce false revelations. To compound the problem, these memories retrieved by leading the subject could still be valid. It's the means by which they were obtained that could be in question. It seemed to me, as we counted down to our scheduled arrival time in New York, that Anna and I were about to test this theory personally. Although neither of us considered ourselves susceptible to leading, we wouldn't know—might never know—until, and if, something was retrieved.

As much as I wanted to discount the cautions we'd heard about hypnotic regression at the M.I.T. conference, I couldn't seem to forget them. Some of the hypnotists and therapists in attendance were blatantly accused of harboring personal beliefs in either good or bad aliens, transferring these beliefs to their hypnotic subjects, thus producing false confirmation of their own views. I had faith in Budd Hopkins and his techniques, not finding his questions in any way leading or suggestive, yet the anxious tickle in the back of my mind remained, perhaps as a necessary safety valve. Only I knew the real truth, however well concealed in my subconscious, and only I could choose to uncover it. As long as I understood the dangers involved, and the possibility of confabulation, I should be able to sort through my mind's protective layers to find the cause for these missing time episodes and eerie fragmented memories. The truth was there and I was going to go look for it.

41. Anna

The weeks before our trip to New York to undergo hypnosis with Budd Hopkins were hectic for me. I resigned, with relief, from my five years of volunteer duties for the horse breed association—

editing a magazine, running an annual horse show, and spending a week in a booth at an international horse show to distribute information. I was hardly functioning most days, and in the last nine months I had pushed myself to the limit, barely meeting my obligations.

My physical and emotional health suffered from my own carelessness and probably from the gray shits' as well. I went to a therapist, but only once. He listened politely as I tearfully related my suspicions about the gray shits and the childhood rape. He then told me that he felt I should enter therapy to fully separate from my father as a normal adult, and to get rid of the psychological torture and terror that this was inflicting on me. He felt he could help, but refused to deal with the aliens. He said he had no experience with possible alien influence and therefore was unwilling to deal with that aspect. No thanks; treat the whole me, not just the piece that's comfortable for you! At $100 per hour, it wasn't worth it. So, I bought some books on parent/child and father/daughter relationships and took charge of my own therapy.

Later, I found another therapist and saw him once a week for several months. He helped me cope with the emotional stress, but knew nothing about the alien abduction phenomenon. He was fascinated, yet skeptical. It didn't matter to him whether he thought what I was relating was real, it was enough that I believed it might be. He was supportive and suggested many techniques, such as automatic writing, to get in touch with my body, my feelings and my subconscious. With his help, I was able to concentrate on getting the daily tasks done—paying bills, doing my work and communicating more with people. Yet it was difficult to cope with my own trauma while educating the therapist. I wasn't strong enough to do that.

I consulted an internist to help me handle the physical problems. I had diarrhea almost every day. Some mornings I would wake up and be nauseous or ravenously hungry. I was exhausted most of the time, and I still had those terrible headaches every day. It was time to get some of these problems fixed. We tried migraine medication for the headaches, we tried beta blockers and vasodilators—all unsuccessfully. I was then given a CT scan of my head to look for any abnormalities that could be causing the constant headaches over my left eye. Since it might also uncover alien implants, it was worth the $1200. The result? Absolutely nothing wrong! I should have been relieved, but I wasn't. I wanted something curable. No implants showed up, so I couldn't blame the aliens either, although I have since learned that the resolution ability of an MRI is more likely than a CT scan to display this small anomaly.

I had a barium enema examination to see if the cause of the bowel problems could be found. Nothing, except a bit of diverticulosis. That definitely was not worth the cost, in money or discomfort! Then the doctor decided that maybe it was all related to stress. "Was I under any stress?" Oh yes, but I didn't tell her about suspected abductions. I wasn't ready to completely believe that myself. I had enough stress with work and family obligations, and financial pressures, to account for my symptoms without bringing in the gray shits. I could accept her diagnosis, for now.

Beth's abductions seemed to be continuing a couple times a month, and this made me uneasy. Was I accompanying her, yet not remembering it? Why won't they let me remember? I'm ready for it. Some of Beth's tales were funny, such as the time she awoke eighteen inches off the bed and then floated down to it and fell asleep. She didn't think it was funny. I guess I didn't feel like laughing much the night I came home from work and Beth told me about her dream. She saw the gray shits inject her with something (she had marks in her hand), and inject me with something in my leg. I looked at my leg where she indicated and found dual punctures!

My own dreams weren't much better. I'd been doing self-hypnosis (meditation, deep relaxation—whatever you want to call it) before sleep to help me remember. A few nights I saw some strange things. One night, I saw a white square with hieroglyphics on it. I saw it twice within that dream. First there would be one symbol, then two, then lots laid down one on top of another. I forgot what any of them were the next morning. In another dream, I saw a huge round disk with scoop marks out of the bottom of it. It was somehow comforting! Several mornings I had woken with a sense of complete and utter peace. Was this feeling also connected to abductions? If it was, my experiences may be very different from Beth's. Her memories were usually terrifying.

One night I saw the hybrid babies—maybe. They looked a lot like the little greys, but were somehow different—and they were only sixteen to eighteen inches tall. Little guys! They didn't scare me, but, of course, over the last few months I'd been telling myself not to be afraid of what I saw. Maybe it was finally working and I was remembering stuff—that, or making it up! I hoped to have some answers after we saw Budd.

42. Beth

The cab dropped us off in front of Budd Hopkins' townhouse. We were right on schedule. Budd welcomed us as old friends, got us settled in his downstairs studio, and generally made us feel quite at home. We would be spending a lot of time in that studio, not only during hypnosis sessions but also in sleep, as the sofa opened into a bed and would serve double duty for three full days.

We spent two pleasant hours getting to know one another, relaxing over drinks and conversation. Budd was a renowned artist and we were impressed with his sculptures and paintings as we roamed around the huge warehouse-like studio admiring his work. This was a side of Budd we hadn't anticipated, and it helped us to see him in an altogether different light. Budd's interpretations apparently ran to the more abstract, yet colorfully and vigorously portrayed, enlivening an otherwise drab and barren space.

Within a few hours Courty arrived at the studio. We had agreed earlier that he could witness and tape the sessions. Courty had previously believed that childhood abuse might be the cause of our troubles. It had taken him some time to conclude that our experiences did not quite fit the pattern. Although certainly some of the traumatic stress abductees were undergoing mirrored those who had suffered abuse as children, the abuses were leveled by non-human beings. And since the abuses were continuing through adulthood, the trauma (often diagnosed as Post Traumatic Stress Disorder) also continued. In fact, we all agreed that under these conditions PTSD was a misnomer; it should be CTSD: Continuous Traumatic Stress Disorder.

Wanting us all to feel relaxed and comfortable with the procedures to follow, Budd took the time to explain the hypnosis process, reiterating a few of the more common misconceptions. Hypnosis, he explained patiently, was not a means of controlling the subject so that one could be coerced by the hypnotist to respond against one's nature. In other words, a hypnotized subject could not be made to do or say anything which he or she wouldn't do or say when *not* hypnotized. Subjects' susceptibility to suggestions varied; some were more suggestible than others. This technique was a powerful tool when used to help stop smoking, lose weight or in some other way alter behavior.

Budd readily admitted that leading questions by the hypnotist could elicit a desired response, of which we were well aware. But, he added firmly, the subject may choose not to be led at all. The downside, though, was that the subject might still confabulate, or even de-liberately lie under hypnosis whether or not leading questions were posed. We hadn't considered this! Did this mean we might be mak-ing all of it up, imagining or fantasizing without knowing it? Yes, Budd said, that was certainly possible in some cases, but when our histories and descriptions of events were taken into consideration as a whole, the chances of fantasy seemed remote.

That made sense, we decided. What kind of person would want to fantasize about being humiliated and physically and mentally abused by frightening, scrawny, *ugly* beings who had complete con-trol? We supposed there might be people out there who were so mixed up, so emotionally disturbed, that the idea held some fascina-tion. After all, sadism and masochism did exist. A person suffering from low self-esteem may even believe the mistreatment was war-ranted. Did we have low self-esteem? Low enough to generate fanta-sies of deserved abuse?

We were scaring ourselves with all this pop psychology stuff! If these events turned out to be imaginary, we could get help; if they turned out to be real, we would learn to deal with them. Either way, we could no longer afford the luxury of denial. It was time to find out one way or the other.

43. Beth's Hypnosis

(Transcripts of these sessions were provided by Courty Bryan. Most of the hypnosis sessions lasted more than two hours, yielding a great deal of detailed information which was impractical to include in its entirety. Therefore, only selected excerpts are used in this and the following chapters.)

Friday afternoon, November 20, 1992

In this first session, I wanted to examine early childhood trau-mas which had given birth to adult phobias. One such phobia was a fear of light, especially overhead lighting. My vivid memories of hid-ing in the bedroom closet would also be explored to determine what, if anything, had triggered the behavior.

Budd spent several minutes helping me enter a relaxed state of hypnosis, though I was certain I couldn't have really been under! I felt alert yet calm, and found I didn't need to concentrate on his voice—yet I heard him clearly. Once Budd was satisfied that a deep state of relaxation had been achieved, I was regressed to an earlier time, going slowly back through the years by imagining myself turning back the pages of a calendar.

Suddenly I was a child again, perhaps four years old, and remembering the house I lived in then; the bedroom I shared with my older sister, Julie; and the reality of being that little girl all over again. I wasn't just remembering what it was like; I was reliving it. I had no knowledge of being an adult, and no sense that there was another place and time from which these images and feelings were brought forth.

Although I did not consciously hear Budd's voice or understand that he was directing me to a specified time or place, I described (in a little girl's voice) what my bedtime routine was like:

Budd: I want you to get the feeling of getting into your bed. Your mommy and daddy come in to tuck you in. Then there's that time when you're alone, with your sister, in that room. It's very quiet now. But there's something that's going to startle you, or frighten you, because we know you went to the closet for safety. We know that already. We don't know what it is, though. Perhaps you had a bad dream; perhaps you heard a noise you can't understand. It could be a lot of things. But everything has a beginning, a first moment when you notice something. When I count to three, that's going to begin. One, you're at the edge of that beginning; two, you're even closer now; *three.*

Beth: [Screams:] *Ayieee! Daddy!* Oh-h-h, Daddy!

Budd: You're safe, Beth. You're here with me. What are you seeing? Why are you calling your daddy?

Beth: The cats are coming in! The cats are back! *Daddy!*

Budd: You're okay. Hold my hand, Beth. Can you feel my hand? Good. What are the cats like. Tell me about the cats.

Beth: They're in the window...an' I can't wake up Julie. I yelled real loud, an', an' my daddy won't come!

Budd: But you see your daddy eventually, don't you?

Beth: Why won't anybody wake up?

Budd: It's okay. This is very upsetting to a little girl, but you're right here with me now. Feel my hand. Tell me what's happening, Beth. Do the cats go away?

Beth: [Screams:] *They're inside the window!*

Budd:	They come in? Are they big cats or little cats?
Beth:	*Big cats! They came in!*
Budd:	We know they finally go away. We know that happens. But let's see now, they're inside the window? Let's just see what's happening now…
Beth:	Uh, it's bright in here. All the light's coming in the window!
Budd:	Is it from the street light outside?
Beth:	No-o-o! There's no light out there! There can't be a light out there! *It's coming in the window and it's going to hurt me!*

Budd attempts to relieve my anxiety by having me play the role of observer rather than participant. He asked me to think about how they made movies, by using cameras that could go up in the air and look down on the scene. Unfortunately, as a four year-old, I had no knowledge of such things and had no idea what he was talking about! Realizing that this tactic would not work, he instead asked me to look at the light and describe its color.

Budd:	Does the light have a color, or is it white?
Beth:	Buh-lew. It's buh-lew.
Budd:	The light's blue? Okay. Now, is there any sound? Do the cats meow or make noises?
Beth:	No.
Budd:	So, what's going to happen with these cats? We know they left, but before they leave, let's see what they do….
Beth:	I don't like to look at them. They're scary! They're putting something in my ear! *[Cries out in pain:]* It hurts!
Budd:	One of the cats hurt your ear? Listen to me, Beth. When I count to three, that ear thing will be all over. One…two…three. Whatever that is that the cat put in your ear, it's all over now. Let's let those cats go away. Do they go back through the window? Is that how they get out?
Beth:	*[In a child's voice:]* I don't know….
Budd:	What happens when they leave?
Beth:	I go hide in the closet, and then I wake up.
Budd:	You wake up in the closet? Did you go to sleep in the closet and then wake up there?
Beth:	No! I don't go to sleep! *[Whispers:]* Gotta watch the door.

Later in the session, as Budd brought me forward to the morning after this event, I described how angry my mother had been when she had to remove all the clothes I'd stuffed against the closet door. She had scolded me for hiding in the closet—yet again—and warned me that she would put a lock on the door the next time I was found there.

I also discovered that I felt a lot of resentment toward both my mother and father, even though I recalled my father being present many times during the "cats" visits to my bedroom. I knew he was aware, and probably helpless to stop it, but I felt abandoned by him nonetheless.

In an effort to obtain a clearer and more coherent description of these cats, Budd brought me back to the present and asked me to describe them from an adult perspective.

"How big were the cats, and what color were they?" he asked.

I saw them as a muted shade of gray and hairless, or very short-haired. They had no ears, I realized, and walked on two legs. I believed the "cats" were about as tall as my father's ribs, which at the time seemed perfectly normal to me!

Hoping I might be able to recall what had transpired before I was discovered in the closet by my mother, Budd asked me look back as an observer and tell him what happened after the "cats" put something in my ear.

Beth: My father's in the doorway. He came. I knew he would come.

Budd: Do you think he saw the cats?

Beth: Yes.

Budd: What happened after you saw your father? Did he take you out of the room, or did he stay with you?

Beth: *[After a long pause:]* He took me out with him. He carried me, like always.

Budd: Where does he carry you? Do you go down the hall or to another room?

Beth: Out, out the *window!* Right *through the window!*

Budd: What's outside the window? Are there trees out there?

Beth: There's a sort of roof on the porch, back porch. And there's a garage over on the right side. We just go between the house and the garage. And there's a big thing *[breath quickens]* and they go in first, and then my father puts me down, and…and, he's crying. *[Beth starts to cry.]*

Budd reassured me that both me and my father were fine and had survived this, that I'd seen my father recently and knew he was okay. Once I was calm again, he asked me to describe where we were and what was happening there.

I recalled seeing several children, some alone and some with their parents, though the adults seemed dazed. I didn't recognize any of these children or adults, but for some reason I felt better knowing we were not alone. Some of the adults (who may have been the children's parents) were nude; this fascinated and embarrassed me at the same time. They didn't seem to know they were naked, or they didn't care.

I described what looked like white tables that were attached to the floor. These tables had no legs and there were no chairs any- where to be seen. As a child, this struck me as odd and I remembered thinking that the place looked like our house just after we moved in and before our furniture arrived.

Abruptly my father and I were separated from this group and led along a suspended walkway with a railing running around the outside edge. Stark white walls curved along the inside of this walk- way, but I didn't remember seeing any doors or corridors in the wall.

I became anxious as I described being led into a large room, also unfurnished except for tall tubes going from floor to ceiling, like glass elevators. I couldn't actually see the ceiling; it looked as if there were none. Still upset, I told of being taken from my father's side and placed inside one of these "elevator things" where I was doused in some sort of fluid. The fluid was neither wet (like water) nor dry (like dust), but somewhere in between. I didn't know the word to describe this material, so I just called it "snowflakes."

Eventually I was allowed to come out and rejoin my father, who appeared very upset; I could see tears running down his cheeks. We were led out of that room through a long tunnel where several of the other children and adults I'd seen earlier were already congregated. Suddenly we were engulfed in a cold fog and I felt myself become very light, almost weightless.

When I could see my surroundings again, my father and I were outside the back door of the house. In a daze, and feeling a little sick to my stomach, I was carried upstairs by my father and placed gently on the floor of my bedroom closet. My sister was still sound asleep and I didn't remember seeing or hearing my mother. My father closed the closet door and I must have then fallen asleep.

I was startled by these revelations, but still couldn't accept them as real memories. It did explain some things; my father's knowledge of my experiences; the unsettling memories of being taken away from my father and his inability to help me; my habit of waking up inside my closet.

What it didn't explain (not to my satisfaction, anyway) was why I had no fear of cats. In fact, I had always liked cats, especially kittens. I told Budd about my fascination with the little stuffed kittens when I was a child and asked him how I could find such things so appealing if these memories were true.

"Well, first of all," Budd ventured, "there are some obvious psychological possibilities. When you saw those things and thought at first they were cats, you were old enough to rationalize that cats probably wouldn't be outside a second-story window, and that cats have fur, and that cats probably couldn't have gotten inside your closed bedroom window—unless you let them in. When you see a real cat, you instinctively recognize that it's not like those cats. Besides," he added, "you can control a real cat."

I laughed at that, and just couldn't resist a little sarcasm: "You can *control* a cat?"

44. Anna's Hypnosis

Friday evening, November 20, 1992

For my first hypnosis session with Budd Hopkins, I wanted to explore the missing time incident in 1990. I explained to Budd what I remembered about the incident while Courty Bryan took notes and recorded the session.

I had been evaluating natural resource work-study programs at southern colleges. Flying in from Nashville, I rented a car and drove to Huntsville, then on down to Tuskegee, Alabama. The missing time occurred between leaving Tuskegee and arriving in Tallahassee, Florida. I left about 4:30 P.M. and arrived about 10:30 P.M. I was told that it would take three and a half to four hours. The problem was that I thought I had to drive all the way to Montgomery, pick up super-highways and then drive back across parts of Florida to get to Tallahassee. But I looked at a map and found a straighter route on a two-lane road. I decided to take that route. I didn't like super-highways anyway. I remember stopping once, looking at a map and thinking, "Am I lost?"

At 8:00 P.M. I was given a speeding ticket just after going through Blakely, Georgia—doing 85 mph! I usually drive a little over the speed limit, but not 30 mph over it! I remember thinking,

"There's nobody on the road, and I'm late. I can't seem to get there from here." When I was stopped by the cop I realized I was disoriented and felt sick, like I had the flu. When he made me get out of the car, I had a hard time walking back to his car. For some reason he wanted me to see the radar detector numbers. I said, "I don't care. Just give me the ticket and let me go on my way." I think he wanted to see if I could walk, if I was drunk. It was 10:30 P.M. when I checked into the motel, roughly six hours from the time I left Tuskegee. I think I may have stopped at a fast-food restaurant on the edge of Tallahassee; I was starving.

Before beginning the regression, Budd and I discussed some of the things that were bothering me:

Budd: So we do have a time problem.

Anna: Unless they were talking super-highways and I took back roads. That's what I assumed at the time.

Budd: Yes, but if you were cruising along at 85, you certainly weren't lingering on the back country roads. "Dawdling" does not spring to the lips as a description of your driving to Florida....

I mentioned a vague sense of a light that didn't seem to be daylight, and that when I checked into the hotel I felt terrible. In fact, as soon as I got there, I took a cold shower—which was an odd thing to do, but I felt so horrible. Budd asked me when I last took a cold shower. I replied, "Never. I was just so hot. And the whole next day I felt weird, half-sick to my stomach." I was convinced I was coming down with the flu; it was hot, the motel was hot. I remember being disoriented: When I pushed the elevator door button, it seemed to open and shut too fast for me to get on it. Finally I pushed my briefcase in to hold the door and smashed the briefcase.

Budd also asked me what else I might like to discuss. I told him I had nightmares as a child, but I couldn't remember what they were about. I slept on the top bunk bed. Every so often my parents would find me on the floor because I had fallen out of bed without waking up. I only hurt myself once; I fell on my sister's shoe buckle. I didn't wake up that time either.

I felt abandoned as a child. I used to cry myself to sleep at night. I was convinced my parents didn't love me. I had no unusual feelings about bedrooms, sisters or closets, though. Our conversation went as follows:

Budd: Any fears or phobias that got focused when you were little?

Anna: Not when I was little.

Budd: What about now?

Anna: Children. Once they get bigger, it's okay.

Budd: If someone handed you a two month-old baby to hold, what would happen?

Anna: I don't think I'd drop it, but I wouldn't do it.

Budd: Couldn't handle it. Okay. When you think about it, some friend hands you a baby to hold wrapped up for whatever reason, to put money into a parking meter. Looking at the baby, what would be the feelings?

Anna: Revulsion.... Fear....

Budd: Obviously, this is a very strong thing.

Anna: *[In tears:]* You're not supposed to feel this way.... It's not normal.

Budd: Did you ever think you were pregnant?

Anna: Only once or twice. I remember getting some home pregnancy tests. Besides last summer.

Budd: No, I'm talking about when you're with somebody.

Anna: Yes, once in high school I thought I was pregnant. My period was a couple of weeks late and we sort of had a celebration when it finally came. And I remember one time, probably ten years ago, and I went through that again and I thought, "Oh my God, I'm pregnant! I can't handle this." And I could have been pregnant, but I wasn't. Then I went and got my tubes tied. I said I'm not even going to take the chance for that anymore. But you know what's weird, I always thought it would be neat to be pregnant, to go through the experience, because I wanted to have the experience. But I didn't want it when it came out. I just thought it would be neat to find out what it would be like.

Budd: You've absolutely never had an abortion?

Anna: Never.

Budd: Because people who say they have ambivalent feelings about a child, that could come from deliberately having an abortion or something like that.

Anna: No, that's very clear that never happened.

Budd: Would you say this fear of children has made you hesitant to get involved?

Anna: I just knew if I ever got involved and married to someone I'd say, "No children. No way."

Budd: So it didn't pull you back from relationships?

Anna: No.

Budd: Let's get specific about the babies, children. Obviously you start from a newborn little baby and range through its really helpless times, then a year old, then it walks, and at two years it starts to talk, and then up through school. Is there any period here where the revulsion is stronger than other periods?

Anna: When they're about that high. [*Holds her hand about three feet up.*]

Budd: So that would be like a year and a half old?

Anna: Walking around size. There's a day-care center in our building and [*with revulsion*] they get on the elevator with me sometimes.

Budd: So "toddler" would be more—how about a new-born?

Anna: [*In tears:*] It's not as bad.

Budd: I'd really like to look into this because when you have something like this that is this powerful, it can cause all kinds of problems in the real world because you're going to run into friends with babies.

Anna: I know. I avoid them.

Budd: You do well with Noel [*Beth's granddaughter*]?

Anna: Noel's bigger.

Budd: When you have that shocking sense of revulsion and fear and everything, do you think of yourself in any way different than you are now? Do you see yourself as another child? When you have that shock and fear, does it flash into anything in your life that suddenly you're a little kid, a twelve year-old girl, a teenager or anything?

Anna: No, in fact, this is sort of weird. When I was going with this guy in high school—we were going to get married and stuff—my parents said, "You go to college, then you get married. Have a life first!" and we broke up, of course. But senior year in high school we were talking about getting married, having a dozen kids, the whole marriage bit and doing everything you're supposed to, and I was okay with that. But sometime in college—

Budd: So you didn't have it when you were little?

Anna: No.

Budd:	It sort of enters at a certain point?
Anna:	After high school.
Budd:	What do you think about when you think of the child being revolting?
Anna:	I've just got to get away.
Budd:	In your worst case scenario, does the child approach you? Ignore you? How does the child behave?
Anna:	The child touches me. Clinging.
Budd:	Again, the worst case scenario is what?
Anna:	It would grab me around the legs.
Budd:	Any other place?
Anna:	Just the legs. I get the feeling that they just want to be all over me.
Budd:	Do several children up the ante of fear more than just one?
Anna:	Yes. If there were like six, I'd.... *[Shudders]*
Budd:	If a child grabbed you around the legs, could you just reach down and pull the child's arms off you? What would you do?
Anna:	I'd run away.
Budd:	But what if the child were hanging on?
Anna:	I'd kick him away.
Budd:	You wouldn't want to touch him....
Anna:	I'd try not to hurt him, but....
Budd:	Is there anything you'd like to ask me?
Anna:	No. I trust you. I'm scared.
Budd:	I know. Just assume at the outset that all sorts of horrible things are going to emerge and say, "Okay, screw it. So we have a lot of horrible stuff and then we have lunch." You're among friends, so if you start out with the assumption, "Okay, what am I going to learn? I'm going to learn I got picked up, and this and that happened." You've read these books....
Anna:	I've read them all.
Budd:	So just imagine that the very worst you've ever read happened. So what? Here you are.
Anna:	Yes, that's what I keep telling myself.
Budd:	Okay.

Before entering the hypnotic state, Budd told me to avoid trying to analyze what takes place under hypnosis, to "be a reporter, not a pundit." I settled onto the couch, was covered with the blanket and made comfortable. My hypnosis began with relaxation and going to a safe place. Ten minutes later, Budd put me into the rose garden for the next few moments, then: "Okay, Anna, I want you to enjoy that beautiful rose garden on a beautiful sunny summer morning. You're feeling so wonderfully peaceful and relaxed...." He put me deeper and deeper: "You're feeling comfortable, relaxed. You're feeling peaceful, relaxed...." And then Budd sets the scene:

Budd: And in this very relaxed state with your mind so alert and so clear, you can see yourself as you were a little over a year ago in the summer when you were on your drive around the south, visiting colleges. Let's go back to Tuskegee. It's getting on late in the day, in the afternoon, and it's time to go. I want you to get the feeling of being in your car now, of getting in your car. Things are packed in the back and you're all set and you start off. It's late afternoon, summer, always interesting and also a little boring just to drive along, but there's some interesting scenery. Driving down it's not too big a road. You get the feeling of the hum of the engine, driving along, heading to Florida, heading to Tallahassee. You're driving along and various things happen on this trip. There are various interruptions, things that were unexpected. We know at one point there's a police officer. So there are some things that happened. I want you to get the feeling of driving on. Tell me what the landscape looks like. You can speak whenever you like.

Anna: *[Softly:]* It's farms. And it's very poor.... It's peanut fields, peanut fields.... There're some flat fields.

Budd: Just look out the window as you drive along.

Anna: Trees. Peach trees....

Budd: As you drive along everything looks kind of typical, kind of boring, but maybe at some point you notice something as you drive along that seems a little different. Just look and see if you notice anything different. It could be any number of things....

Anna: *[After a pause:]* It's just boring.

Budd: Boring? Uh-hm. Just keep looking as you drive along.... We know for sure there was at least one surprise on this trip. We know that for sure. If there's one surprise, there could be more surprises. Something that breaks the monotony and the expectations. I'm going to count to three now, and my hand is going to come down on your hand and it will feel safe and it will feel nice. At the count of three we know there's at least one surprise. We'll go to the very first moment. When I count to three, you'll get that first inkling. One, you're driving along.... Something's going to interrupt the monotony. Two, right on the edge now.... Three!

Anna: *[Long silence]*

Budd: Tell me, what are the little inklings? You might sense something different....

Anna: It's like a crown of light.... *[Whispering:]* Fire. Like a gas stove burner.

Budd: Like a gas stove burner, um-hm.... Is this off in a field somewhere?

Anna: *[With wonder:]* Yes....

Budd: Where is it in the field?

Anna: I don't know; it's on my right.

Budd: Just keep looking at it.

Anna: Gone.

Budd: It's gone? Do you mean you passed it by?

Anna: No, it just went away.

Budd: Is there any traffic on the road? Other people might notice things....

Anna: No. I'm all alone.

Budd: Do you see road signs, where you are?

Anna: Stop sign. I turn right....

Budd: You're just driving along, let's just see what you see. Just report what you're seeing. A little town?

Anna: No. Nut trees.

Budd: How're you feeling?

Anna: Pretty good.

Budd: Bored, maybe?

Anna: Yes, it's a long drive.

Budd: A long drive. And you're driving along.... How's the car performing? Are you having any car trouble?

Anna: It's okay.... I have a feeling it's stopped.

Budd: What's stopped?

Anna: The car. I can't see it. I just feel it....

Budd: Did it stop with a jerk, or did it just slow down gradually?

Anna: Gradually.

Budd: And all of your senses are extremely alert. Your sense of feeling—

Anna: Cold!

Budd: You're cold? Is the air conditioner on in the car?

Anna: I don't think so.

Budd: Tell me what you're feeling. Your body is very, very aware of anything it feels. It felt a little cold....

Anna: Waiting.

Budd: Waiting? And how are you waiting?

Anna: Sitting there.

Budd: Waiting always means a period to which there's going to be an end. Whoever or whatever you're waiting for is going to show up, or you're just going to leave. What's going to happen now?

Anna: *[Sharp breath]*

Budd: Tell me what's happening?

Anna: *[Surprised:]* There's somebody at the window!

Budd: At the window of your car? Um-hm. Is this on the driver's side?

Anna: Yes.

Budd: Okay. Here's the thing: A quick little glimpse. When I count to three, just through your eyelids, a quick little look, and see who's there. One, getting ready to take a quick look.... Two.... Three! A quick look and then close your eyes. Who's there at the window?

Anna: I don't know! ...It's a long face.

Budd: A long face?

Anna: *[Getting tense:]* It's not a person!

Budd: Tell me about the person.

Anna: *[Shakes head: No.]* I'm supposed to go with him.

Budd: Does he tell you that?

Anna: No. I just know.

Budd: Now this is the south. You could have been seeing a lot of black people. Is this a black person?

Anna:	No. That one's gray.
Budd:	Gray. Right. Do you lock the door of the car?
Anna:	No.
Budd:	What do you do?
Anna:	I get out…. Why am I doing this?
Budd:	Let's worry about that later. When you get out, do you stand next to this person?
Anna:	Yes.
Budd:	Is he a big person?
Anna:	Yes…. But we sort of float away.
Budd:	Let's allow that feeling, allow that feeling to happen, that floating.
Anna:	That's weird. It's, it's nice!
Budd:	Does he touch you when you float or just—
Anna:	No.
Budd:	So you're floating. Where are you floating to?
Anna:	I don't know…. I'm upright.
Budd:	Upright? Like a foot or so above the ground?
Anna:	Yes.
Budd:	Does this seem like a totally new experience, or does it seem familiar?
Anna:	It seems neat!
Budd:	Okay, let's see where you float to.
Anna:	Octagon! Sharp sides. There are lights there.
Budd:	On the thing that looked like a gas light?
Anna:	Yes. Like a gas burner…. But it's like a flange down there.
Budd:	You can take a good look at it and make a drawing later for me. Just make your eyes a camera and look at it. So, what happens next?
Anna:	*[Quietly:]* We go straight up. Inside….
Budd:	What's happening?
Anna:	Just standing there.
Budd:	Where are you standing?
Anna:	White room.
Budd:	What's happening?
Anna:	Squiggles. Squiggles.

Budd: What do you mean, "squiggles"?

Anna: On the wall or something. Squiggles. Like on a black screen.

Budd: Now I'd like you to take a very good look at those squiggles for me. Get a sense of them so they're still in your memory, so you can draw those squiggles for me later on.

[Note: After the session I drew what I had seen. After Budd looked at my drawing, he brought out a notebook and showed me drawings made by other abductees. Some of the figures were exactly the same!]

Budd: Okay. Are you alone in this room, or is the person with you who took you in?

Anna: Yes, sort of. Eyes.

Budd: What?

Anna: Eyes. More eyes....

Budd: Where do you see these eyes now?

Anna: Around.

Budd: Um-hm. A lot of people standing around looking at you?

Anna: Shorter eyes.... Yes, there're three or four of them. There's one over there [points to left side], and there're two over there [points over left shoulder].

Budd: Okay, we're going to get a sense of what they're going to do. Did anybody tell you what this was all about? Do you ask them what's happening?

Anna: "Why'd you do this?"

Budd: What did they say?

Anna: "We need you." [Whispering:] Why? [Exasperated and sad:] "You don't need to know!"

Budd: That's what they say? "You don't need to know?"

Anna: Yes. [Begins to cry]

Budd: That's not very forthcoming, is it?

Anna: No.

Budd: Did they ask your permission for this?

Anna: No! It was neat in the beginning, but.... [Exhales sharply:] It's something else now! "Can I go home now?"

Budd: You want to go home now; you mean back to the car?

Anna: Wherever. They don't think so. So I can't.

Budd: Is that something you feel, or something they say?

Anna: I don't know....

Budd: What's happening? Are you still standing there?

Anna: I'm lying down.

Budd: Okay, this is what I want you to do now while whatever this all is [is still going on]. I just want you to keep your eyes closed and not to look at anybody. But I want you to know, Anna, that your body is extremely sensitive; it has its own memories, its own memories that can feel, for instance, the surface that you're lying on, what that surface feels like. Your body can feel the sense of the fabric, whatever's next to your skin, clothing, a t-shirt, whatever you're wearing. What are you wearing?

Anna: Nothing. It's cold! *[Shivers, pulls the blanket up to her chin]*

Budd: We'll put this up over you here; we'll just warm you up. You'll feel much, much warmer. Your body remembers exactly what it's feeling, what it felt. You're lying there, and you're cold, and you don't have anything on. This is what we're going to do. We're going to start with your feet and we're going to move up from your feet systematically through your whole body and see what your body's memories are. We don't know. But we're going to see. We're going to start with your feet and it's cold.

Budd asked me about my feet, ankles, legs and thighs. I didn't feel anything unusual, except for a large bruise on my left thigh where I remembered a horse kicking me a few weeks prior to my leaving on this trip. It hurt all over again under hypnosis!

Budd: Painful. Okay. Now being very systematic, you're lying there and this is going on, moving up now to your female parts, to your genitals. What do you feel in that area that feels different in any way? Or does that part of your body feel normal?

Anna: Tight.... *[Pauses, then says almost wonderingly:]* Cramps.... Oooh.... Owwww!

[I was lying on my right side, facing Budd. I began to writhe in pain and drew my knees up in response to severe cramps.]

Budd: More specifically, is this your abdomen, or is it down lower?

Anna: Down lower.

Budd: In your genital area?

Anna: *[In obvious discomfort:]* Yes-s-s. Cramps!

Budd: Now your body has very, very good memories and can sense what's happening, can sense what's causing this sort of feeling. What are you sensing in that part of your body?

Anna: Pain…. *[The pain mounts:]* Pain! Oh! Oh! Oh-h-h!

Budd: Pain. Inside you?

Anna: *[Writhing:]* Ohhhhh.

[Budd turns down the pain by using an imaginary rheostat.]

Anna: *[With obvious relief:]* Whew!

Budd: Now when this is happening, is there any movement involved or is it still? What does it seem like? What would cause something like this? If you could duplicate—

Anna: It's like everything was just squeezed into a little tiny knot!

Budd: As if there's pressure again from the outside?

Anna: Yes…. Outside, but inside, too.

Budd: Is it steady, or is it intermittent?

Anna: It comes and goes.

Budd: Is it connected to a feeling of some kind of movement? Or is it just a still presence that's affecting things?

Anna: *[Sudden, horrified whisper:]* Oh, my God!

Budd: *[Worried:]* Is it happening again?

Anna: Ohhhh. *[Breaks into tears. Does not answer.]*

Budd: Tell me what you're feeling now. You're okay, my hand's on your head now. Tell me like a good, clear reporter. Tell me what's happening.

Anna: *[In tears, voice breaking:]* I feel like I just had a baby!

Budd: You feel like you just had a baby?

Anna: I wasn't pregnant!

Budd: You're okay now. What was that feeling like? Did you feel that something passed through you?

Anna: Yes-s-s.

Budd: Something came from the inside and went out?

Anna: Yes-s-s. The pain's gone.

Budd: Good. Now, just to quickly—

Anna: *[Disbelief:]* That can't be!

[Note: Beth, at this point, is in tears.]

Budd: Don't worry about that. Let's not even worry. We don't know what this experience is, and we're not going to try to guess. This is what I want you to do: Probably at the time you opened your eyes just a little bit to just glance and see what's happening. When you glance down, do you see anything?

Anna: It's tiny. It's very tiny.

Budd: How tiny is it?

Anna: About the size of a pear.... Oh-h-h.

Budd: And what happens to this pearlike thing you're seeing?

Anna: *[Matter-of-factly:]* They're taking it away.

Budd: Do you at any time look at it as its being taken away? Can you see what it seems to resemble?

Anna: It's pink and wrinkled.

Budd: Does it have appendages? I mean, you feel like you had a baby. Does it look like it has—

Anna: It's very short. Real short.... Little, tiny.

Budd: Do you hear any tiny cries from the—

Anna: No.

Budd: Okay. So it's taken out. Now, how do you feel? Do you feel much better that it's gone?

Anna: *[With relief:]* Yes.

Budd: Okay. Good.... Do they say anything to you about this?

Anna: It is their's.

Budd: It's their's. Do they say that to you?

Anna: Yes.

Budd: Do you ask how this came about? What do you say to them?

Anna: "How can it be? Where did it come from?"

Budd: Now this is what I want you to do for me right now. Let's just float slowly away from that experience because your body is extremely sensitive with its own memories. Let's go back some months before this. If you were pregnant at some point—we don't know whether you were—but if you were, your body is going to know that somehow a seed was inserted into it somehow to produce that baby. Maybe you felt absolutely nothing, but maybe you felt something. So right now, I want you to concentrate all your attention down to the sexual parts of your body. I'm going to count to three and in those months or even weeks before this happens, your body is going to remember if it ever felt anything connected with this. One, concentrating your attention down to that part of your body…. Two, right on the edge…. Three….

Anna: *[Without hesitation:]* Tube. A long tube like Dr. Tom uses.

[Note: This refers to the artificial insemination tube that my veterinarian uses on the mares.]

Budd: A long tube?

Anna: Yes.

Budd: Where are you when this tube is being used?

Anna: I don't know…. Someplace white.

Budd: Are you by yourself, or is there a doctor there with you?

Anna: No. Just them.

Budd: Just them.

Anna: *[Surprised:]* Ohhh. Why are they doing this?

Budd: Does it hurt?

Anna: It's weird.

Budd: What's the diameter of this tube, if you start with the diameter of a pencil. Is it that diameter or is it bigger?

Anna: Littler.

Budd: Littler. Thinner?

Anna: *[In wonder:]* It's flexible…. Clear.

Budd: Clear. Is it attached to something on the other end?

Anna: Like a syringe.

Budd: A syringe. Okay.

Anna: White stuff in the syringe…. I guess it's sperm.

Budd: Here's the thing: Memories are extremely clear. This is what I want you to do: Somehow or other you were brought to that place and this was done to you. I want you to back up in time as if you were backing up a movie. I want you to see where you were before they took you to that place. Let's move back. Where are you?

Anna: I'm in a field. Pulling weeds!

Budd: What's happening?

[Note: Both sets of tapes containing the next portion of this session were lost. This portion of the transcript is reconstructed from notes taken by Courty Bryan.]

Anna: I'm pulling weeds in the field. It's summer. The weeds are low.

Budd: Where is this field?

Anna: It's the log cabin field [*a field on the farm in which the ruins of an old log cabin remain*]. They're in the field. It's a little space ship.

Budd: How big is it?

Anna: About the same size as a station wagon. About twenty feet long. They walk up to me. I have to go with them.

Budd: Do you try to run away?

Anna: No. [*Smiling*] We're floating toward the ship. It's neat! There's a ramp with steps.

Budd: Who is with you?

Anna: There are three of them. One on each side and one in the front.

I was brought into the craft which appeared to contain just one large, round room. In the middle of the room was a table.

Anna: I have to get on the table.... [*Worriedly, as if talking to someone:*] "I don't think I want to do this anymore. Can't we just not do this anymore?"

Budd: What's happening?

Anna: Oh, I'm tired. They won't let me walk out!

[I was evidently laid out on the table, naked.]

Anna: They've got that tube again!

Budd: Is it like the tube before?

Anna: Yes. They put it in me!

Budd: How far?

Anna: About six inches. It goes into the fallopian tubes. Ow, it hurts! [*Curls up in pain*]

Budd established that the tube was inserted into my vagina, past the cervix, and into my left fallopian tube.

Budd: Have they done this to you before?

Anna: Yes! [*Bursts into tears*]

Budd:		How old were you when they did it the first time?

Anna:		[*Outraged and hurt:*] I was just a kid. I was twelve! "Oh, God I hate you. Leave me alone! Go choose somebody else!"

Budd:		Where were you when this happened?

Anna:		It was at night when I was fishing with my father. On a canal.

Budd:		Who else was there?

Anna:		It was just me and my dad.

Budd:		What is your dad doing?

Anna:		[*Angrily:*] He's just standing there!

Budd:		Does he try to help you?

Anna:		He can't. He can't protect me.

Budd:		What is he doing?

Anna:		He's scared. He's just standing there. He can't move.

Budd:		How do you feel about your father?

Anna:		I feel hatred for him! He can't protect me!

Budd explained that there was nothing my father could have done, that he had been paralyzed by them, and that instead of hating my father, I should direct my anger at them. He asked what the aliens were doing to me.

Anna:		They hurt me.

Budd:		Where do they hurt you?

Anna:		It hurts me inside. It's too big! I'm just a little kid!

Budd:		Where is it inside?

Anna:		It's in the vagina and it's big. It hurts!

Budd:		Tell me what you're feeling.

Anna:		It's like a tearing sensation.... I'm bleeding from my vagina.

[Note: It is established that this whole "operation" took place in front of my father on the banks of the canal.]

Anna:		It's like we never left that place. They did it right there.

Budd:		Did your father see all this?

Anna:		I don't think he could see. It was done in front of him, but I don't know if his eyes were open or closed.

Budd took me back to the Alabama-Florida incident and asked what happened after they had taken the pear-sized baby away from me.

Anna: After they took the baby away, they cleaned it up and me up.

Budd: Do they show you the baby?

Anna: *[Shudders:]* I don't want to see it!

Budd: Why don't you want to see it?

Anna: I hate babies! *[Disgusted:]* They're not human! They don't look right.

Budd: Have you seen these babies at other times?

Anna: Oh, yes!

Budd: Tell me what they look like.

Anna: They're not babies. They're just walking around. Thirty or forty of them. They have big heads. Fatter bodies than the gray guys. Some of them have little bits of blond hair.

Budd: Do they say anything? Do they speak to you?

Anna: They make noises.

Budd: What sort of noises?

Anna: They squeak.

Budd: Why have they brought you to all these babies?

Anna: They want me to pick them up…. *[Makes face, revolted:]* Ugh. Someone gives me one! I have to hold it! I don't want to do this! I almost dropped it.

I admitted later that I did drop the baby, but was ashamed to tell Budd that—even under hypnosis. The whole encounter with the babies was so distressing that Budd decided to take me away from that scene to my childhood in England. I recalled meeting a little girl with glasses and blonde curly hair.

Anna: She asked me what England is like. I tell her it's neat. The people talk funny, but they have ponies all over.

Budd: Does she tell you her name?

Anna: I don't think we tell each other our names.

Budd: Look around yourself. Tell me where you are. What do you see?

Anna: We're talking outside some place.

[At this point I don't know where I am or how I got there. Budd told me that my body remembered, and to let my body's memories tell me how I arrived.]

Anna: I floated there from a ship.

Budd: What was the little girl wearing?

Anna: She has a blue, red and white striped shirt.... No, it was either blue and white, or red and white.

Budd: How old is she?

Anna: She's older than me. She's in the 6th grade, I'm in the 4th. She tells me she has a gray dog.

Budd: Do you see her dog?

Anna: It's not there.

Budd: Have you ever seen this little girl before?

Anna: We know each other.

Budd: How do you know each other?

Anna: We met when we were littler.

Budd: How little?

Anna: I was in a playpen. She's outside. I'm inside.

Budd: Where is the playpen?

Anna: It's inside something.... A bedroom.... There's a single bed. She doesn't have her glasses on.

End Transcript

After the hypnosis session, Beth and I discussed this incident and I drew a picture of her bedroom in Maryland. According to Beth, my drawing was a good representation of the bedroom she had from age four until she was about nine, when she moved from Maryland to Virginia.

The debriefing after the hypnosis session allowed me to make some sense out of what my subconscious just revealed. We also talked about past counseling which focused on what I had thought was the problem, the alleged rape at age twelve by my father. Now that it was revealed as an alien encounter experience, I'd hoped the trauma would fade.

"The proof is in the detoxification of the trauma," Budd told me. The alleged rape was not the problem; that wasn't what was causing the trauma because it was not the source of the trauma. "It's not a hidden memory," he added. "You were being treated for the wrong disease." The proof would be if there was a sense of relief. The headaches would ease off; depression would ease off. "There will be a sense of wholeness." Budd told me that everything I had said and observed about the rape was appropriate to the age. Decades of mistreatment had to be undone little by little.

I should have been relieved, but I wasn't. The new memory was too fresh, the old one too long-standing.

Later that evening, I remarked to Budd that "I must have made the whole thing up." His reaction was immediate: "You're an awfully good actress, aren't you?" No, I had to admit, I wasn't. The emotions and the pain that I felt were so real. If it weren't for the emotions, it would have been much easier to believe I'd extracted this from my reading—but I'd not read anything like this.

Yet I must have made it all up. This was so confusing! I did admit to Budd, "I guess I can't deny my contacts with the little gray shits anymore."

While relaxing after dinner in a bar with a rock and roll band, I started crying uncontrollably. Courty and I went for a walk and I admitted to him that I had been completely overcome by feelings for another lost baby. I knew they had taken another baby from me while I was in college. The songs from the seventies had brought out a grief that I never knew existed within me, let alone acknowledged.

I cried for my lost children.

45. Beth's Second Hypnosis

Saturday evening, November 21, 1992

For this second session, I asked Budd if we might examine the dreamlike memory of July 15, 1992, in which Anna also appeared. Anna was asked to go upstairs (out of hearing range) during the session since Budd did not want her to be influenced by what I might recall under hypnosis. Anna would be hypnotized later to determine how she remembered the incident.

Once under hypnosis, I was regressed to the night in question:

Beth: I'm calling Cricket *[one of their two dogs]* to come back into the house. *[Calls:]* Cricket! *[Sudden, sharp exhale:]* Whew! Must be a plane.... No, its—it's no plane.... Oh-h-h, neat! It's one of them! *Anna! Anna!* You won't believe this! *[Impatiently:]* Hurry up! Look at this! *[Pause]* You know what I think? I think it's one of *those.* Look how *big* it is! It's *huge!* It's like a fish...

[I believed, as I relived this incident of seeing a UFO at close range, that this was the first time I had ever seen anything like it. I soon discovered that this reaction was quite common, and that I seldom recognized either the aliens or their craft until I was well into the experience.]

Budd: Where is Anna? Is Anna with you now?

Beth: Yes. We're both on the porch. This thing is so bright! It has a piece at the end like a fish tail…. *Oh-h-h.*

Budd: What do you see, Beth?

Beth: It's kind of wagging at us. Hurry up! *[Angrily:]* Hurry! *[More quietly:]* Anna doesn't want to go. We're going to miss it.

Budd: What is it doing?

Beth: It's waiting for us down there. *Nobody will believe this!* *[Laughs]* So? Wipe it off!

Budd: What happened?

Beth: Anna stepped in manure! She told me to watch out for the manure, and *she* stepped in it! *[Impatiently:]* Hurry up! Come on! Look at it up there!

Budd: What is Anna saying, Beth?

Beth: She says it's cold. It's not *that* cold!

Budd: What's happening now, Beth?

Beth: We have to go over there. And Anna was supposed to come with me. We're supposed to do all these things together.

Budd: How long have you had to do these things together?

Beth: *[Extremely anxious:]* I don't like this!

Budd: How long have you had to do these things together, Beth?

[I suddenly reverted to childhood, remembering other incidents where Anna and I had been together. Although this was an unexpected turn of events, Budd permitted me to pick my own way, not knowing what might be uncovered.]

Beth: *We were babies!* We were just little kids!

Budd: You and Anna?

Beth: I was holding her up. She could hardly sit up. We have to be together, take care of each other.

Budd:	How old is Anna?
Beth:	She doesn't talk. She's just scared. She doesn't know what they are.
Budd:	What *who* are, Beth?
Beth:	They're on the floor. Like a tile floor, on a blanket.
Budd:	Are these other children?
Beth:	[*Nods, yes.*]
Budd:	What do these children look like?
Beth:	They don't have any color. They're not gray and they're not white.... They're...they aren't any color.
Budd:	How old are you, Beth?
Beth:	[*Suddenly lapsing into a little girl's voice:*] I'm four. *She's* just a baby!
Budd:	Are there any adults around?
Beth:	There's a lady...
Budd:	What does she look like?
Beth:	She's just a little person. She has funny hair.
Budd:	Funny hair? What kind of funny hair?
Beth:	Stringy. Like she's going to go bald.
Budd:	What color is her hair?
Beth:	Yellow.
Beth:	[*Whispers:*] She's scared. *Sh-h-h!* She wants her mommy.

In my mind, this little girl (who was barely old enough to sit up by herself), was helpless without me to care for her. When Budd asked me where the baby's parents were, I told him her parents weren't there and that was why she was so frightened; she missed her mommy.

When Budd asked me how we had gotten to this place without our parents, I told him "we flew there."

"Where were your parents?" he prodded. "Weren't you with your mommy and daddy?"

I responded that I didn't know, but that I was sometimes in the closet when "they" came to get me, so I couldn't see if my parents were there.

Budd:	And you're going to be her mommy right now?
Beth:	[*Whispers:*] Yes.
Budd:	What color hair does the little girl have?
Beth:	She has yellow hair, too.

Budd:	Just like you do. So, at any rate, you calm her down. Now, let's see what's happening. So there she is. This lady with the funny hair. Does she say anything to you?
Beth:	She doesn't talk.
Budd:	What's she wearing, this lady? Does she have a dress on?
Beth:	No. She, she has…loose skin. It's funny skin. She wears a thing over here—[indicates above the left breast] like a pin?
Budd:	What's on the pin?
Beth:	I don't know what it is. Maybe it's her name.
Budd:	Does it have writing on it, looks like writing?
Beth:	No…. There's something there.
Budd:	Does she tell you you're going to have to take care of this little girl?
Beth:	Yes.

Since there seemed to be little else I could remember about this incident, Budd suggested I move to the next time I saw this same little girl and describe the circumstances.

I immediately cried out, complaining that "the little brat" had just pinched me. I described myself as the long-suffering older child (about seven or eight), and the one who pinched me as a blonde girl about five years of age who I had seen many times during our "flying trips." I called the little girl A.J., explaining to Budd that this was a "nickname" I had given her.

I recalled spending a good deal of time chasing after A.J. because she was constantly running off. I was often nervous, I told Budd, when she wouldn't listen to me, because I believed her antics would get us into trouble with the "babysitter." The "babysitter" was the same one (or very like) the stringy-haired, pale-skinned female who was usually present.

But this time I saw several younger versions of the "babysitter" in the same room; I had trouble describing them because I didn't recall ever seeing anything like them before. They were not childlike, but sat motionless most of the time, as if by moving they might crumble. These "children" were fragile looking; most were bald but some had a few white hairs dangling over their eyes, which were oddly pale and lifeless. These tiny replicas of the "babysitter" showed no interest in the structured games A.J. and I were asked to play, and usually just sat perfectly still, like rag dolls. (See Figure 5.)

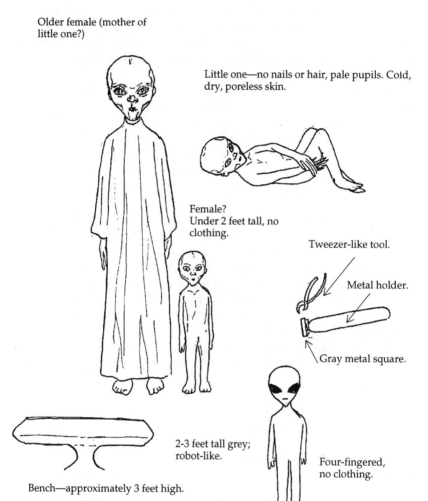

Older female (mother of little one?)

Little one—no nails or hair, pale pupils. Cold, dry, poreless skin.

Female?
Under 2 feet tall, no clothing.

Tweezer-like tool.

Metal holder.

Gray metal square.

2-3 feet tall grey; robot-like.

Four-fingered, no clothing.

Bench—approximately 3 feet high.

Figure 5. Fragments of Beth's abduction memory from January 1992.

When I could manage to get A.J.'s attention, I would try to do what was asked of us. A.J., though, considered these "games" boring. One such game involved manipulating a square gray box (which was sealed shut on all sides) so that it would be smaller than it first appeared. This miraculous feat had to be accomplished without touching the box! After a long time spent concentrating on the problem (and holding A.J.'s attention), we were finally able to turn the box *inside-out!* The result was that the box was now actually *smaller*, just as we were told it would be.

Another game required that we move a stick (protruding from the wall) up and down in that wall, again without touching it. We were to "think" the stick down the wall to within reach, grab hold of it, then "think" it back up again while hanging on for dear life! I let A.J. do this one first while I stood beneath her in case she fell. This was much easier than we expected and after a while we began to enjoy the game, asking that we be allowed to play with the "stick in the wall" again.

Yet another game, one we both feared, pitted us against one of the taller greys who pitched a "fire stick" at us, instructing us to throw it back at him. Each time we attempted to catch the stick, it would emit a horrible electrical shock and burn our hands. Finally, in anger and frustration, we joined forces against him by concentrating on "catching" the stick and "throwing" it back to him without actually touching it. After several tries, we succeeded, hurling the "fire stick" back to our tormentor, ending the game once and for all.

I wasn't sure whether these activities all occurred during the same abduction experience—or over several—but I suspected we were asked to do them more than once.

Moving forward in time, Budd asked me to imagine myself as a fifteen or sixteen year-old, in high school.

Budd: Do you see A.J. in your high school?

Beth: *[In a more mature voice:]* ...No.

Budd: When's the last time you see her? When's she's little?

I described seeing A.J. in a department store in Washington, D.C. There was another girl with her and an older woman walking between them. Hiding behind a clothes rack, I tried to get her attention by whispering, "A.J.," several times, but she didn't seem to hear me. I had been playing on the floor with one of my Dee-Dee Kittens, moving in and around the racks, and was for some reason afraid to be seen by the people with Anna.

After describing this scene, I tried to remember the next time I saw her:

Beth:	I think that's her. I think so.
Budd:	And where do you see her?
Beth:	*[Exasperated:]* Oh-h-h!
Budd:	What happened?
Beth:	Bugs!
Budd:	Bugs flying around? Are you outside?
Beth:	Yeah.... Geez, it's hot!
Budd:	Where are you?
Beth:	Woods.
Budd:	In the woods. Uh-huh.
Beth:	There's a field out here.... Where the hell is everybody, anyway?
Budd:	So what are you now, about thirteen or fourteen?
Beth:	No, I'm, I'm twelve years old. I'm going to be thirteen pretty soon. We're on a school picnic at the park, but I don't want to be with the other kids. God, I hate bugs!
Budd:	You're going to be thirteen pretty soon. You're out in the woods and those bugs are flying around.
Beth:	Besides, I hate this, anyway. I hate this, this wandering around looking for A.J....
Budd:	Looking for A.J. again. Do you have to do that often?
Beth:	She's always off somewhere. I can never find her when I need her.
Budd:	I see. Let's watch. I want you to see exactly the way A.J. comes, when you first see her.
Beth:	*[Makes little lip sounds, puh-puh-puh, like an exasperated twelve year-old]*
Budd:	Maybe she just comes walking up through the woods. I want you to look and see where you see A.J. first.
Beth:	I want to sit down and wait for her.
Budd:	You know you're going to see her?
Beth:	Yeah, she's supposed to be here. She's in trouble again, I betcha.
Budd:	Let's watch for the very first minute you see her. Tell me how she comes.
Beth:	*[Surprised:]* Oh! Don't do that...! The witch!
Budd:	What did she do?
Beth:	*[Irritated:]* She comes up behind me!

Budd: So you didn't see her come?

Beth: *[Impatiently:]* I don't know where she came from. She comes up—why does she always do this? What?

Budd: What does she say?

Beth: *[Impatiently:]* Where have you been? *[Almost singing:]* You better be care-ful-l-l. Don't do anything to make them mad.

Budd: *[Apparently misunderstanding:]* We don't want you to get mad at her.

Beth: If you do anything, then it makes it worse. I know what I'm talking about. Listen to me!

Budd: What does she say to you this time?

Beth: She thinks it would be neat. It's not neat! *[Pause, then quietly:]* They're letting her hold those babies…. And they're not neat, I'm telling you.

Budd: What about the babies? I didn't understand this. She said what about the babies?

Beth: They let her hold the babies.

Budd: They let A.J. hold the babies?

Beth: Yeah. Dumb, Anna! I'm telling you, don't let them do that. Don't let them do that! They start letting you hold babies; I'm telling you, don't let them do that!

Budd: Did they do that to you?

Beth: Oh, yeah, lots of times they did that.

Budd: You had to hold the babies?

Beth: Yeah. I didn't like that either.

Budd: What do the babies feel like? I bet they squirm around and cry, don't they—

Beth: *[Impatiently:]* No, they don't! They don't do—it's like holding something dead! Disgusting!

Budd: —squirm around?

Beth: No, they don't. They're just like those other little tiny ones that were always there. They don't do anything. They're like they're dead. I don't even think they're real! I don't know what they are, but they're not really like babies.

Budd: Now, when you hold those babies—

Beth: *God!*

Budd: —are they heavy?

Beth:	No. They don't weigh anything. They're like paper.
Budd:	Like paper, uh-huh. Now, when you hold them, how are you supposed to hold them? On your shoulder?
Beth:	No, you're supposed to hold them...
Budd:	In your lap?
Beth:	No. You're supposed to hold them up close, and walk them. And one time they told me I was supposed to let the baby nurse.... I didn't want to do that!
Budd:	Did you think you had breasts that a baby could nurse at?
Beth:	I know I could. I've done that before.
Budd:	You know you could. Did you have milk for the babies?
Beth:	*[Whispers:]* Yes.
Budd:	How did you know?
Beth:	Yeah, I know it was there. I know all about that.
Budd:	It was there. Who taught you all about that?
Beth:	*[Whispers:]* I don't like to talk about that.
Budd:	Who taught you about these things?
Beth:	*[No response]*
Budd:	Did someone teach you about these things?
Beth:	*[Unintelligible]*
Budd:	Was it a bad boy who did something to you the first time?
Beth:	*[Crying:]* It wasn't fair! He just didn't know, maybe.
Budd:	Was this a boy you went to school with?
Beth:	*[Quietly:]* No.
Budd:	Where did you meet him?
Beth:	In my bedroom.
Budd:	In a bedroom? Where—
Beth:	*[Stiffens suddenly, sharp intake of breath]*
Budd:	It's okay, it's okay.
Beth:	*[Panicky:]* He's very cold! *He's very cold!* Don't—
Budd:	I'm going to put this blanket over you—
Beth:	Tell him...tell him he's cold!
Budd:	Let's pull the blanket over you, okay? Where are you when he does something? Just tell me where it is and who it is.

Beth: *[Crying, frightened:]* In a bedroom, a bedroom.

Budd: Who is he? Look at him. Let's just get a—

Beth: *[Firmly:]* I don't want to look at him! I know what he looks like.

Budd: What does he look like?

Beth: *[In tears:] I know who he is!* I've seen him before. Lots of times. *I know who he is! [Panting:]* Unhh! He does things! And he always tells me, "Everything is fine. Everything is fine." And it's not fine, because—*[starts kicking and thrashing around on the sofa]* Go away from me! *Go away!*

Budd: Is he doing something to you?

Beth: He's right on top of me! *Go away!*

Budd: Let me hold your hand. This is me holding your hand. Feel my hand? Okay. What does he do to you?

Beth: *[Does not answer]*

Budd: Is he somebody you know?

Beth: No. I mean, I know who he is, but I don't know…

Budd: Does he have a name?

Beth: No, he doesn't have a name.

Budd: What does he look like?

Beth: He looks like *them!* All of them!

Budd: What color hair does he have?

Beth: He doesn't—have hair. He doesn't have hair anywhere. He's like rubber. He's cold all the time, and he touches me, he's cold. *Don't do that!*

Budd: What's he doing to you? Now I want you to take—Beth, take a deep breath. Just rest a minute. Okay. Are you warmer?

Beth: It hurts.

Budd: What hurts? What part hurts?

Beth: Down there.

Budd: Down between your legs it hurts? Is that where it hurts? Is it down in your female parts, in that part there is where it hurts? What did he do? Just tell me what he did.

Beth: *[Wearily:]* He made another baby.

Budd: How does he do that? Does he have a penis he puts in you?

Beth: No…no….

Budd:	He doesn't do that. What does he do?
Beth:	He gets over top of me and he puts something he has, he put it in down there and gets on top of me, and it just all goes in there. And then he takes that thing out and then it doesn't hurt anymore. *[Resigned:]* It's no big deal.
Budd:	What I want you to do right now, just as if he's right in front of you this minute, tell him what you think about this. Tell him what you think about it….
Beth:	I don't want to look at him.
Budd:	Let's look at him indirectly. What would you like to say to him right now? You can say whatever you feel like saying.
Beth:	*[Very quietly:]* I don't think I should do that.
Budd:	*[Very firmly:]* Beth, this is *your* body and *your* mind. You can do what you want to—
Beth:	*[Crying:]* They can do this anytime they want to! They can come and do stuff. They can come and put things in and take things out. *They can do anything they want!*
Budd:	Let's speak to him directly. What if he says to you, "I have your permission"?
Beth:	*[Starts thrashing and pounding the sofa furiously]*
Budd:	Okay, okay. Take a deep breath—a deep breath now, Beth, take a real deep breath. You know he's lying when he says that. Nobody has the right to do that to you. Will you listen to me for a minute? Nobody has the right to do this to you; no one has the right to do anything to you that you don't give them permission for. And you didn't give him permission, and that's not right. And you have every reason in the world to be angry.
Beth:	I fixed it.
Budd:	You fixed it.
Beth:	I fixed it.
Budd:	They can't do it now, can they?
Beth:	*[After a pause:]* They can't ever do that again.

[I had, in fact, removed myself from that time completely, and was remembering when I'd had the hysterectomy and how relieved I'd felt. I had rationalized that the greys would never do that to me again because I could no longer get pregnant.]

Budd:	Okay. Feeling a little better? Take a deep breath. You fixed them! Do you have a right to your own body, Beth?

Beth: It doesn't hurt anymore.

Budd: Tell me, do you have a right to your own body?

Beth: *[Softly:]* Yes.

Budd: You have a little granddaughter right now.

Beth: She's so cute.

Budd: If she said to you, "Grandma, people are trying to do things to me, should I let them?" What would you say?

Beth: *[Crying:]* I don't know *what* I'd say! *You can't stop them!*

Budd: Wouldn't you say, "It's okay to try and resist them"?

Beth: *[Crying:]* Yes!

Budd: Try to resist them.

Beth: I don't know *how-w-w!*

Budd: Yes, but you say, "Try to resist." That little girl has a right to her own body, her own life, doesn't she? She does have that right. And you're going to be able to tell her and help her with that. See, you had nobody to help you, did you, when you were little?

Beth: My daddy tried.

Budd: Your daddy tried, but she has somebody who can really help her, and that's you. You can be so helpful to her! You can talk to her on the phone, write her, and when she comes to see you, you can be very helpful. And you're going to feel good about that. Now I want you to take a very deep relaxing breath.... Just relax, just relax.... Let's go back to that other little girl, A.J., and the times you saw her. When was the first time you saw her when you were a woman, an adult?

Beth: I don't know if it's her....

Budd: Of course it's confusing. Here's what we'll do. So we're now getting up to the present. Let's move back to that field where you were very exasperated because Anna was so slow, and she wouldn't come, and she stepped in the manure....

Beth: *[Laughing:]* That's Anna for you!

Budd: And there's something about—there's something you don't particularly want to talk about.

Beth: Oh, boy! Do we have to do this again?

Budd: Yes. Let's just see what that was about.

Beth: *[Nervously:]* I don't know if I want to do this....

Budd:	Well, you told me a lot about this dream. There's some-thing about—there is something about....soldiers? What's this about soldiers?
Beth:	They're not soldiers.
Budd:	What are they?
Beth:	Guards.
Budd:	How do you meet these guards?
Beth:	At the door.
Budd:	What's the door for?
Beth:	It's in the tunnel.
Budd:	Where's this tunnel?
Beth:	*[Sharply:]* Underground! We're underground. I don't like this. This is terrible! It's cold as shit down here! *Oh!*
Budd:	How'd you get there, underground?
Beth:	I don't know. We're down underground. We're down here and Anna's not happy. She is definitely not a happy camper!
Budd:	What's happening underground?
Beth:	We have to go into this big room. It's like a hanger or something. Big, huge hanger. (See Figure 6.) *[Sighs, voice shakes:]* Oh-h-h, boy! Wow! I don't think Anna thinks this is neat at all. Oh, well.... A guy puts something on me.
Budd:	What's that he puts on you?
Beth:	Some machine or something; it makes my ear ring, my right ear. There's a bunch of people over in the corner.
Budd:	Let's look at the people. What do they look like?
Beth:	Lot's of people. I know a lot of them. I think Anna does, too.
Budd:	Let's look at those people and see who they are....
Beth:	Hoh-h-h-h, there's Gail! I haven't seen her in a hundred years! I don't remember her last name. I knew her a long time ago, went to school with her. And there's Dee....
Budd:	She used to be your best friend?
Beth:	She was my best friend.
Budd:	Does she recognize you?
Beth:	Yeah, and her husband's with her, too. I'm surprised he's here.
Budd:	What's he wearing? A business suit?

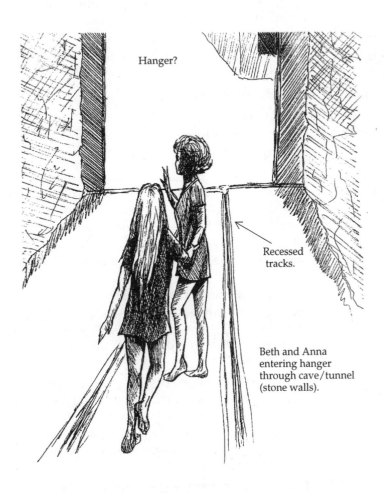

Hanger?

Recessed
tracks.

Beth and Anna
entering hanger
through cave/tunnel
(stone walls).

*Figure 6. The beginning of a joint abduction experience in July 1992,
first thought to be a shared dream.*

Beth: No, he's in just a short-sleeved blue shirt and a pair of
slacks.

Budd: What's Dee wearing?

Beth: A red dress. Very pretty.... Wow! This is great! This is
wild, why're you here?

Budd: Like a reunion? Do you know everybody there or just
some of them?

Beth:	No. I don't know everybody.
Budd:	Let's look around and see if there's anybody else....

I named off a few other people both Anna and I knew, including Leslie and her two children. Leslie boarded her horse with us and had previously mentioned having weird dreams and missing time episodes.

Beth:	Anna's talking to somebody over there; I don't know who those people are. She's talking to two people over at the side.
Budd:	Let me just ask you, is Courty there?
Beth:	No.
Budd:	Am I there?
Beth:	No.
Budd:	Is there anyone else there you recognize?
Beth:	Yes, Kelly. She's not talking to anyone though.
Budd:	Just seeing different people....?
Beth:	There's a lady over there that I met before; I don't re-member where. And there's some people from the barn. Some people are upset; some people are scared and con-fused.
Budd:	What do the chairs look like that people are sitting on?
Beth:	Not too many people are sitting down.... They're metal folding chairs.... Oh! *[Quickly:]* Here they come! What are these guys? *[Referring to "guard's" helmets or masks:]* Whoa! I want one of those to ride in, then I wouldn't get arena dust in my eyes! That would be great!
Budd:	What is that?
Beth:	It's a helmet, or something. Covers the whole head and face.... Geez, can you breathe in there? *[Louder:]* Hey! *[To herself:]* I guess so...okay. Then they're going to take some people. They pull out these things and point them at people and then they pick some and they go, over there to this thing. Oh! Stuff is falling out of the roof. *The sky is falling!*
Budd:	There has to be a reason why you're all together here.
Beth:	I don't know. There's something out there in the middle of the floor. I know what it is, but I don't believe it. It's probably not real. I know, it's a dream or something. And they close the door now.... *Bang!* Whew, heavy-duty door! There are more people going. *Big* machine!

This thing is really big, and there's stuff falling all over it. *Ping-ping-plunk-ping!* When those things open up, it makes a rumbling noise and everything just *shakes!* Grass and rocks and stuff are coming down. What does this mean? What does this mean? This means something. I'm supposed to know what this means. Oh, well…. We get to go now. I don't really want to go in there.

Budd: Go in where?

Beth: I guess I do want to go. I'm curious. I really want to see what's in there. Besides, it's just a dream, a dream. I, I just wonder…. I don't know if I get to come back. *[Stronger:]* Nobody came out of it! They went in there and they didn't come out! I didn't see them—

Budd: Who went in there?

Beth: Lots of people. They went in groups, sucked right up, like somebody sucked them up through a straw, but you couldn't see the straw.

Budd: They don't walk up steps?

Beth: No. There aren't any steps I can see. I see all these people I know and I try to talk to them, but I'm afraid I'll miss something. Something's going to happen. We're going to get to go now. Look happy! It's just a great dream! *[Softly:]* A dream….

Budd: Where are you going?

Beth: This thing is sitting up on a crane? Something's holding it up at an angle. There's these humongous caterpillar things, with a driver's cab on top of it. (See Figure 7.) You could put a man in there and you wouldn't even be able to see him! Nothing is this big! This is a great dream!

Budd: It sounds like a terrific dream. So, you go up like the others do—sucked up?

Beth: I'm looking up at these ball-like things underneath it. Wow! Look at that! I wonder if they light up? I think I've changed my mind about this….

Budd: You have a choice.

Beth: Really, I can—I can walk away from here. I don't have to do this. This is just, this is just a dream. I don't have to do this.

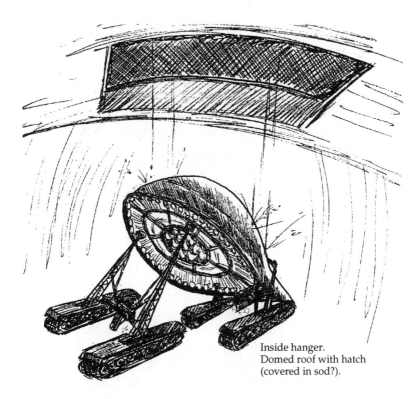

Inside hanger.
Domed roof with hatch
(covered in sod?).

Figure 7. The alien craft seen by Beth during the July 1992 joint abduction.

Budd: Do you want to go away from it?

Beth: I'm so curious, though. Why do I have to be so damn curious? Oh-h-h, crap! I'm going up inside it. Oh-h-h, this is going to make my stomach turn over, I know it. I should never have had dinner.

Budd: Where are you now?

Beth: I'm inside the thing.

Budd: What's it look like inside?

Beth: This place is absolutely *huge!* It's like in layers and you can look up through sections, and there's a whole other section up there, probably another one on top of that. It's like a *city* in here! Wow! This is a really big place! There's lots of people in here....

Budd: What percentage of the people do you think you know?

Beth: I don't see Anna anywhere. I see some others I know, about six or seven people.

Budd: Do you get that flash of recognition on somebody's face when they see you?

Beth: No, we just don't do anything. We look at each other. Nobody seems real sure about this. Um, we're moving, we're moving. I hate this, because it never feels like you're going anywhere, but you are....

Budd: Do you decide where you want to move to?

Beth: No, I think they're just sending us places. It's like shipping the mail, you know?

Budd: Where are they sending you to?

Beth: Down to this place that's like a round hall. It's got lights on the side...and tubes. And we're going down single-file. I really shouldn't have eaten.

Budd: How's everybody dressed? Is everybody different?

Beth: Yeah.

Budd: Does everybody have clothes on?

Beth: Everybody has something on—not very much for some. One man only had his jockey shorts on. That's funny; he's too tall to be that thin, but he has real long muscles—very wiry. *[Laughs]* He has, he has a scar on his leg.

Budd: Where's his scar?

Beth: On his thigh. His left thigh? *[Sarcastically:]* I wonder where he got that from?

I went on to describe a long tunnel we passed through that was shaped like a semicircle and seemed to descend gradually. I also noticed an obnoxious odor, but couldn't think what it reminded me of.

Budd: Is it like a chemical smell, or something different?

Beth: Sort of like—I don't know. It doesn't smell like anything, it smells like lots of things. Like something burning, like burned skin, like you burned your dinner. Yuck!

Budd: Do you see anybody else you recognize now?

Beth: That woman, she looks like Ellen, but she has a different name. I call her Katherine.

Budd: What's she wearing?

Beth: She has on a t-shirt, but it's too short and she's not wearing anything underneath it. The t-shirt's got something

on the front of it, in big red letters, like "crap" or something equally crude. I can't really read it. Some of the
stuff has worn off, pieces of the lettering, or decals.

[Note: We discovered later that Ellen, who was a member of
Budd's support group, had a favorite nightshirt with a worn and
peeling red decal on it.]

Budd: When you get down there, what happens?

Beth: Over on one side there're these trays, flat trays. They're
like trays, only they're not quite trays.... They're
like...uh...metal and they have handles on one side.
There are eight of them over on the side there. We're
standing up against where the wall curves up to the tunnel thing we came through. There's stuff on this one tray
over here.... Oh, that looks familiar. There's this itty-bit-
ty scissor-like thing on the end of it. It's in a tube thing
that compresses it, and two little clawlike things come
out and hold something. Oh! I know what that is! It's the
thing they use in the ears! I didn't come all this way for
this! *[Laughs]* Golden opportunity, right? We're going
down past them. I keep trying to look at the tray, but after a while I can't see it anymore. There're other things on
the tray. I can't tell what they are. We go past tables...and
I know what those are, too.

Budd: Is there anything on the tables?

Beth: *[Agitated:]* No, but I know what they are. This is turning
into a nightmare, isn't it? I've seen those tables before.
And there's those box things. They have stuff in the boxes. *[Speaking very quickly:]* There are things coming out of
the boxes that have these long hoses on them—stuff they
put down your stomach. And those things that come
down from the ceiling. I hate these places! *[With relief:]*
We're going on by. We're going up this incline that's like
a corridor that spirals up to other levels. And we go into
this other room. *[Worriedly:]* I don't want to go in here....

Budd: What's happening?

Beth: We're going up to the babies. They have baby everythings here; they have baby babies, baby horses, baby
kangaroos, baby mice, baby everythings!

Budd: Are there adult animals and people, too?

Beth: None of those! It's all babies. Baby here, baby there, baby
baby everywhere. Oh, God, I like babies, okay? I really

do! But you just can't have everybody's babies. Enough with the babies already! God, I don't want anymore!

Budd: Would you like me to bring you back home now?

Beth: *[With relief:]* Yeah.

End Transcript

I should have slept like the dead, but the images from that earlier hypnosis session kept swimming frantically around in my head, keeping me awake. Could any of it be true? Could it really have happened? Or was it just a wonderfully vivid dream that turned a little sour toward the end?

It felt real. It felt, in my gut, like a memory of something that really happened to me. I could still feel my stomach heaving in response to that awful smell; I could still see those trays, those instruments, or whatever they were. It was all so very convincing! But I didn't want to believe any of it.

Besides, what about all that other crap that came out? Where did that come from? How could I have experienced those things and not remembered them before? Had Budd led me? I didn't think so. Courty didn't think so. Budd didn't think so.

Certainly there was something very strange going on. Did I have some secret desire to be a part of this mysterious phenomenon? I didn't think so. Were this the case, I would have chosen a more powerful role to play! I would have been in charge! I'd never enjoyed being manipulated by others. I was an independent and gutsy child, and felt I was a self-assured adult.

Except when it came to this.

Well, Anna would be going through her "similar" dream tomorrow. I doubted anything substantial would come from it. After all, people *do* experience similar dreams. It can't be all that unusual. Perhaps it's just not often reported.

I rolled over and glanced at Anna's sleeping form. I felt guilty for having drawn her into this mess. If I had left well enough alone, things might have sorted themselves out. Now we were going to have to see it through. I wondered if I was up to it....

46. Anna's Second Hypnosis

Sunday, November 22, 1992

When Courty arrived at Budd's studio on Sunday morning, I was in tears, feeling overwhelmed by guilt over having blamed my father for the "rape" when I was twelve, thirty-two years before. I was still not entirely convinced that it was the aliens. (One image has not yet replaced the other; instead, there is a new image, a new scenario *and* the old one.)

I remembered that after the rape my father never understood why I wouldn't go fishing alone with him again. At about the same time, I developed a terror of land crabs; I wouldn't get out of the car if land crabs were around. I'd always thought it was just a child's excuse not to go fishing. Another possible reason why is that the land crabs' eyes, like those of the alien rapists, were wide apart and very prominent.

We decided to hypnotize me concerning Beth's "dream" of a UFO in the underground hanger and possible childhood linkages. The importance of this, according to Budd, was in the "architecture of the acts: Beth and I might have known each other as children and acted as magnets for one another." Budd thought it was possible the aliens planned friendships, relationships and brought people together. But what did this mean in terms of relationships *and* control? I do not believe they are capable of complete control of our lives and our minds. Abductees have at least as much control as pets, but how much of my life have I determined? How much of my actions were directed by them? The aliens, Budd argued, could control the body, but not the mind. An example: They want you to hold the baby, but you seem to have a choice. I argued that induced memory loss seemed to be a pretty good start at mind control. The aliens seemed to do that fairly well, although how much memory loss is due to them and how much to my own subconscious protecting me from trauma is unknowable.

After the relaxation techniques and induction of hypnosis, Budd asked me to go back to July of 1992:

Budd: There's going to come a certain night, however, back in July, a particular night, you're upstairs getting ready for bed and on that particular night you're going to hear Beth call to you. Tell me what her voice sounds like, what she's saying to you.

Anna: Scared....

Budd: And what do you do?

Anna: I run downstairs and out to the porch. Lights!

Budd: You see lights?

Anna: *[Puzzled:]* Upside-down blue ring.

Budd: What did you say?

Anna: Like a gas burner.

Budd: The lights look like an upside-down gas burner…. Where do you see this upside-down gas burner?

Anna: Between the chicken coop and the ash tree, over the orchard.

Budd: What's Beth saying to you? Let's listen to what she's saying.

Anna: We're supposed to go. She's got my hand. She's leading me off the porch….

Budd: Tell me what's happening. Where are you going?

Anna: Back towards the trees. I think it's overhead and we're following.

Budd: Are you excited and eager to follow this?

Anna: Yeah…. It's an adventure. It's exciting…. Ow-w-w!

Budd: What?

Anna: I stepped on something.

Budd: What kind of shoes do you have on?

Anna: I don't have any shoes on.

Budd: What did you step on?

Anna: Yuck! Something sort of squishy.

Budd: What do you think it is?

Anna: I hope it's mud! I don't think it was though, I'll tell you.

Budd: Do you stop and clean it off?

Anna: No, I scraped it off with a stick. Beth's leading. She's got my hand; it's okay.

Budd: Do you have a flashlight with you so you can see where you're going?

Anna: No.

Budd: Is the moon out?

Anna: No, but it's bright enough to see.

Budd: Okay. *[After a long pause:]* What's happening?

Anna: *[Surprised:]* A mine shaft! Timbers!

Budd:	So this mine shaft—
Anna:	Tracks! Goes down. Like for a railroad car….
Budd:	Why are you going down this mine shaft?
Anna:	I don't know. Beth's leading.
Budd:	Is it tall enough to stand up in, or do you have to stoop?
Anna:	Beth's stooping. I can walk upright.
Budd:	I bet it's kind of scary isn't it, to go down the mine shaft at night like that with no flashlight?
Anna:	No.
Budd:	How do you see in this mine shaft?
Anna:	The walls are green. And phosphorescent.
Budd:	What are you seeing? What's happening? I bet it's a small, tight little place, isn't it?
Anna:	No. Big! Clay floor…. How can they get clay this far underground?
Budd:	I bet it's really dark.
Anna:	No, there's lots of light.
Budd:	What kind of light fixtures are they?
Anna:	I don't think there are any. It's like…two or three aircraft hangers.
Budd:	Sounds very big.
Anna:	It's got rock on the wall, faced with rock. Clay floor and rock on the wall.
Budd:	Is the rock rough like they blasted into rock? Or is it—
Anna:	No, it's sort of smooth.
Budd:	Just look around.
Anna:	It's weird. It's like there's astro-turf down here. There's people and folding chairs, brown steel folding chairs. But nobody's sitting down.
Budd:	Let's look at the people. See anybody you know, for instance? Just strangers? Tell me about the people.
Anna:	Most of them are strangers. Some people we know are there. Leslie and her two little kids are there.
Budd:	What's the expression on their faces? Are they glad to see you?
Anna:	Relieved. They know somebody.
Budd:	What about the kids?

Anna: Kids are just hanging on to Leslie's hands. They're just sort of there…. *[Excited:]* Ohh! Jenny is here!

Budd: Let's look at Jenny. Have you seen Jenny since then?

Anna: *[Crying:]* No!

Budd: How do you know Jenny?

Anna: We were in school together. In college. I haven't seen her in twenty years! There's a man with her. He has dark hair…. How'd she get here?

Budd: What's she wearing?

Anna: A gray sweatshirt and black pants or dark blue.

Budd: How many people are down here, would you say?

Anna: Oh, a hundred. Lots! Lots. Big groups of people.

Budd: Okay, let's look around and see if you see anybody else you know.

Anna: Mark. What's he doing there? *[Short laugh]* Sure has lost his hair.

Budd: What's Mark wearing?

Anna: Flannel shirt and jeans. Red flannel. Checks.

Budd: Let's look around and see if you recognize anybody else. You might see somebody else.

Anna: So many people.

Budd: Okay. Now, are you still standing next to Beth holding her hand?

Anna: Uh-uh, no. She's on the other side talking to somebody.

Budd: How far away is she? Ten feet?

Anna: No, fifty, a hundred feet.

Budd: You think she has a different view of things?

Anna: Yeah.

Budd: Now let's look through the crowd. Is everybody dressed in sort of casual clothes, or is somebody dressed differently?

Anna: There's some person in a business suit. He doesn't belong there.

Budd: How about anybody without any clothes on?

Anna: No, most everybody's got clothes on.

Budd: Anybody have like pajamas on or something? You have a nightgown, right?

Anna: Yeah. Lots of little kids in pajamas….

Budd: Any adults?

Anna: Most of them have real clothes on.

Budd: Now, are people talking to each other? Do you talk to people? Do you talk to Jenny, for instance?

Anna: Yes.

Budd: What does Jenny say she's been doing since you've seen her all these years?

Anna: She says she's a child's doctor, a pediatrician.

Budd: Did you know that before? Or is that something—

Anna: When I knew her she wasn't even in med school.

Budd: Does she have any children herself? Do you talk about that? I assume the man she's with is her husband, or is that somebody else?

Anna: I don't know.

Budd: You haven't talked with him?

Anna: He's just with her.... I don't know if she has any kids.

Budd: Did she tell you where she's living?

Anna: California....

Budd: Did she say what town she's living in, in California?

Anna: Sacramento.

[Note: I was able to track Jenny down some time later, and this information proved to be false.]

Budd: Did she ask about you? What you're doing?

Anna: We didn't really have much time to talk.... Everybody's talking now.

Budd then asked me to look around the room and see if there was anyone else there I recognized. Budd offered me names of people who were abductees as well as other people we both knew. I did not see any of those people.

Budd: I want you to look to see if there's anybody practically undressed.

Anna: *[Surprised:]* Yes! He's got jockey shorts on. Hairy chest. He's tall. He's got hair on his shoulders, too.

Budd: Have you seen him before or is he a stranger?

Anna: Don't know.... Hmm, he's got slippers on his feet. That's weird.

Budd: Let's look down his body since he doesn't seem to have much on. Does he have any other distinguishing features besides the hairy shoulders?

Anna: He's got a big scar on the inside of his left calf, a long cut. It's all puckered like it's a real bad scar about six inches long. His legs are kind of skinny.

Budd: Does he look muscular, well-built?

Anna: Not well-built, but he's muscular. Strong.

Budd: How old a man would you say he is?

Anna: Forty-five.

Budd: So he's just in his jockey shorts? Does he seem embarrassed?

Anna: No. There's a light shining down....

Budd: Let me ask one thing: Is this the first time you've ever been in this place? Does it seem familiar in any way? Or is it all new?

Anna: It seems familiar.... Different people, though.

Budd: You think you've been here more than once?

Anna: Yes.

Budd: You said there's a light shining down from up above, from the roof? Or where is it?

Anna: It looks like an opening. Like sunshine's coming in. Like a sunbeam.

Budd: What else is in this room? We have the lights, we have the people, we have the rocks around the side, we have astro-turf, the folding chairs. Is there anything else in this room?

Anna: There's something coming out of the walls. Like half an arch? Like a runner in the middle as though something's supposed to slide on it? It comes out of the wall. Goes up to the floor.... I don't know what it's for. It looks like it's trying to hold the wall up. Like a support.... It's dark in the corners. It seems to go way back. All black.

Budd: Now, you come in there and you and Beth kind of separated from each other a little bit, right? You said you were some distance apart. Now let's just see what's going to happen next.

Anna: Most people are gone.

Budd: Where'd they go? Did you see them go someplace?

Anna: No....

Budd:	Do you see Beth? Is she right there with you?
Anna:	Yes.
Budd:	So, it's just you and Beth and everybody else is gone?
Anna:	Yeah, there are six to eight other people around. Jenny is gone…. *[With wonder:]* A green elevator shaft with eyes?
Budd:	Before we get into that, let's move back a bit because at some point you're going to see what happens to those people. We know they leave because they're not there.
Anna:	Have to line up. Have to line up…in groups. We march away.
Budd:	Each person?
Anna:	No, a whole line of people.
Budd:	Where are they marching?
Anna:	To the left, around the edge, towards that arch thing.
Budd:	Is it possible that you march over there, too, since the other people are? Just see if you march over there.
Anna:	No…. We march toward that elevator thing…. It's behind the arch.
Budd:	So you go over to that elevator thing. Let's just see where you go now. Are you and Beth together or are you separate?
Anna:	No, we're together. Umm, weird. It's like we go into a black hole.
Budd:	We don't know what this is, but tell me what you see in this black hole thing. Any light?
Anna:	We can see each other, but we're sort of…tumbling and floating. We're going down, but it's sort of like feathers floating down. There! We end straight up on the bottom.
Budd:	So you go down, is this straight down? Is this a straight drop, or like an incline?
Anna:	Like an incline.
Budd:	Are you in line, or by yourselves?
Anna:	Nope, there's just us. We're with Leslie and Jan and their kids.
Budd:	Since you moved on, it's obviously for some purpose. Let's see what you move on to.
Anna:	There's a…a "Close Encounter" ship.
Budd:	What do you mean by a "Close Encounter ship"?

Anna: It looks like a big one. I don't how they got it underground, my goodness!

Budd: Is it resting on the ground? Is it hovering? What is its situation?

Anna: It's hovering. It doesn't make any noise. A few lights flashing. Except it's not as big as the one in the movie.

Budd: And what happens when you go into this room and this big thing is there? Incidentally, I didn't ask, but since we all know about these experiences, do you see any of those people around you associate with the UFOs, that aren't human?

Anna: There are some over by the ship now. Someone led off the groups, but they weren't weird.

Budd: What did they look like?

Anna: Like people.

Budd: Now, sometimes people who are ushering have like a flashlight or some kind of thing to point directions. Do they carry anything or are they just empty-handed?

Anna: I don't think they're carrying anything. They just say, "Follow me," and we go.

Budd: Let's look at the face of the one who led you. What's he look like?

Anna: He's got a pointed chin—oh, shit! He does sort of have those eyes…. Not real bad.

Budd: So once you're down there, let's see what happens when you go down this incline thing, whatever that was.

Anna: Mm-mm, we're just sort of milling around…. There's some sort of urn. Gold. We have to put our hand on it. Top of it has this round ball…. For identification? Fingerprints? It's warm.

Budd: It feels warm. Is the room cold?

Anna: No….

Budd: So, if somebody were there in his jockey shorts, he wouldn't be cold?

Anna: I'm not cold. Underground it's supposed to be cold. I'm not cold.

Budd: How about your other senses. We've done sight, but your sense of smell, your sense of hearing, we've had the sense of touch. Are there any other senses that come in here? Do you smell anything? Hear anything?

Anna: Clangs. Metal clangs. Echoes.... I start to get scared.

Budd: Whatever it is, we know one thing, Anna, that you came back fine. When this is over, you're okay. What's happening? How about sense of smell, anything there? Sometimes air can be stale underground.

Anna: Acrid. Could just be me. Smells like vinegar. Just a little odor.

Budd: Can you just walk up to the ship? Bang on it with your fist if you wanted to?

Anna: I can't reach it. It's too tall. Might just be able to touch the bottom of it.

Budd: That's what I meant. Do you think it's close enough to reach if you walked over to it?

Anna: Maybe on the bottom I could. It's too tall to touch it on the edge.... A door! Ramp. Whoops! There we go!

Budd: How are you going? Slow? Fast?

Anna: Floating. Fast. Just sort of picks you up and you go. Except you don't lose your balance.

Budd: So you float up fast into this.... Is Beth with you? Do you see her, too, going up?

Anna: I don't see her anymore.

Budd: Okay, so when you float up, what happens? When you get up there, what do you see? Where are you?

Anna: Spiral-like thing. A spring? There're portholes and there are stars out there! This is dumb! Can't be! Crab Nebula? How can that be?

Budd: Why do you say Crab Nebula?

Anna: That's supposedly what's there to the right.

Budd: Is that something you recognize, or is that something they tell you, or what?

Anna: They told me that. Oh, shit, Jo's here.

Budd: Jo's there? What's Jo wearing?

Anna: He's gray. He's one of the shits. Why's he telling me this?

Budd: Do you think he's telling you the truth?

Anna: Nope!

Budd: So you don't believe him. So let's see what's going to happen next.

Anna: Somebody came in with Spock-ears [the Vulcan alien played by Leonard Nimoy in the *Star Trek* series] and a

pointy, larger head. They're almost like bat ears. They're
cute, but he's ugly. (See Figure 8.)

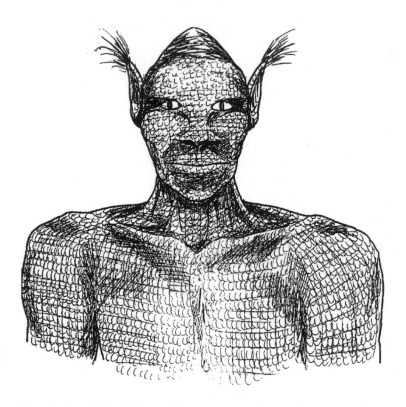

*Figure 8. "Spock-Ears," the reptilian creature seen by Anna during the July 1990
abduction.*

Budd: Now who's in there besides you, and Jo, and the man
with the Spock ears?

Anna: Nobody.

Budd: What kind of room are you in? What kind of space? Big
space? Little space?

Anna: I'm around in the front side where the stars are.... Something behind me, to my left...controls...box.... Spock-
ears is there. There's something to it.

Budd: Is Spock-ears tall?

Anna:	Yes. He's taller than Jo. Whoo! He's quite tall! He just stood up. Oh! Cute little ears though.
Budd:	Why do you think they have you in that room with them?
Anna:	They want to show me something….

I saw what appeared to be a female pangolin (an anteater). I couldn't figure out why they had a pangolin in the control room.

Anna:	I don't want to sit down in that chair. But I have to…. "Okay!"
Budd:	Is this one of those metal folding chairs?
Anna:	No, it's a soft white chair. Shit, I can't get up! Little balls.
Budd:	There's what?
Anna:	They're like light balls. *[Sighs again]* "Do anything you want, I don't care." I'll just sit here. Not going to do anything."
Budd:	Is your nightgown long enough to keep warming you?
Anna:	It's gone. *[Whispers:]* God, I want it on right side-out this time. Dumb shits.
Budd:	Do you want it right side-out because sometimes it's put on wrong side-out, is that it?
Anna:	They're not too bright that way. "Can I go home now? I'm bored. I don't care." *[Impatiently:]* "You know, guys, this is getting pretty old! If you're not going to do anything, let me go home. I got to go to work tomorrow! I need some sleep. Get on with the show, or send me home."
Budd:	Are these little lights still there, bouncing around?
Anna:	Nope, they're gone.
Budd:	Do you think these things were alive, or were they just lights?
Anna:	They just were.
Budd:	So do they let you go home now? Or do they have something else in mind?
Anna:	Yeah, I think I've got to sit there for a while.
Budd:	How do you feel sitting there with no clothes on?
Anna:	I don't give a shit.
Budd:	Well, let's see the next thing. Let's move this ahead to the next event. It may just be that you get up and walk away, we don't know.

Anna:	Clouds.
Budd:	Clouds?
Anna:	Volcano? Oceans? Whoa, we're going fast!
Budd:	Where do you see these things?
Anna:	Weird, but true…. Just floating along.
Budd:	Now, let's look out of the corner of your eye and let's see if you see any of them. The people.
Anna:	Yep.
Budd:	Where are they?
Anna:	One on each side.
Budd:	What are they doing?
Anna:	They've got my hands…. God, this is like Superman! Zoom!
Budd:	Are the men looking in your direction or away from you?
Anna:	They're looking ahead. I can sort of look around, though. I mean, I get to look down. It's okay. It's sort of neat.
Budd:	Do you see cities? Roads?
Anna:	Cities, yeah, New York City. I've never seen it like this, though. They say it's New York City, though. Why do we have to land here when I don't have any clothes on!
Budd:	Where are you landing?
Anna:	Somewhere in New York.
Budd:	Is it nighttime?
Anna:	Yes.
Budd:	Where do you land?
Anna:	Lots of buildings….
Budd:	Do you land at a big building? On a street? Or at a little building, or what?
Anna:	A street outside a big building.
Budd:	People notice you're standing there with no clothes on?
Anna:	Nobody around. *[Impatiently:]* "Why'd you do this, you dumb shits?"
Budd:	Let's see if you see any landmarks. You might see a street sign, or a number, a sign on the building….
Anna:	There's a dumbbell-shaped sort of glass on the building.

[Note: I did see a street sign—Oak—but I thought it was too common a name to mention.]

Budd: Is there a number on it?

Anna: Twenty-one. Oh, we're going into the building. Whew!

Budd: How do you get into the building? Standing up in the door?

Anna: Nope. The door didn't open.

Budd: Is it a glass door? Wooden door?

Anna: Glass. Big. It's like a bank inside.

Budd: Counter like a bank has with tellers and all that?

Anna: No, like a counter where you can get deposit slips and stuff. Some of the chain things, those velvet things. Oh, it's all carpeted and there're steps going up. It's red carpet. Maybe it's a theater. I think it's a theater.

Budd: Anybody in this building? Besides the three of you?

Anna: No-o-o. There's a stage down there. We walk down towards the stage.... Well, we don't walk, we zoom. *[makes floating motion with hand]* That's what I mean.

Budd: Is it dark there, or are there lights on?

Anna: It's dark except for the stage. "I don't have time to sit through a play. C'mon guys. What is this?"

Budd: Are there hidden lights behind the curtain shining down? What do you see?

Anna: Just up on the top. There's only one set. Between the curtains. Little short curtains. Supposed to meet somebody here.

Budd: Let's move ahead to that. Do you recognize who you meet, or is this a stranger?

Anna: It's a stranger. He has a hat on, he has a coat on.... God, he's got clothes on! This isn't fair! He doesn't care.

Budd: Does he—

Anna: "I thought I looked pretty good, too!"

Budd: And what does he say to you? What made you say that just now?

Anna: He looked at me.

Budd: He looked at you like he was attracted to you, do you mean?

Anna: *[Nods, yes.]* God, he's good-looking. "I don't want to go away now."

Budd: You don't want to what?

Anna: "I don't want to go away now! C'mon!"

Budd: Who's telling you that you have to go away now?

Anna: The gray shits are taking me away.

Budd: Why don't you want to go away now?

Anna: He's nice. I'd like to get to know him.

Budd: Does he speak to you? What's he say his name is?

Anna: Brien Schultz. (See Figure 9.)

Budd: Do you know whether he's from New York or where he lives? Does he say where he lives?

Figure 9. Brien Shultz, a human man brought together with Anna on several occasions.

Anna: No.

Budd: Does he have a mustache, beard, any facial hair?

Anna: He has sandy-brown hair. Almost blond, but not quite.

Budd: Does he look at you—after all, you're not dressed—does he look at you as if he's thinking erotic thoughts?

Anna: Yep. That's okay. Yes. Shit! Another married one. Goddamn wedding ring on.

Budd: He's married, uh-huh. Ever seen him before?

Anna: Don't think so.

Budd: Can you imagine what he might look like when he was sixteen?

Anna: *[Disappointed:]* Oh-h-h, I thought you were going to say naked.

Budd: Well, let's do that.

Anna: Yes. Sort of nice. Ummm.

Budd: Do you see him naked, or are you just guessing?

Anna: Just guessing.

Budd: Does he touch you?

Anna: He gives me a hug. Tells me it's okay to go. We'll meet again…. How does he know that?

Budd: Think of him as a sixteen year-old. Do you have a picture of him in your mind?

Anna: Yes, he has light blond hair, he's got freckles!

Budd: But you're putting him as having freckles now.

Anna: He had freckles when he was sixteen.

Budd: Does this seem just like a guess or are you pretty sure about it?

Anna: I'm pretty sure, because he has a rugby shirt on. Green and white.

Budd: Where are you?

Anna: *[Without hesitation:]* England.

Budd: England. Let's look at that. Is that outdoors or indoors?

Anna: Outdoors.

Budd: Does he have a British accent?

Anna: Not Cockney. Not a real strong British accent. Continental, whatever that means.

Budd: When he speaks in the theater, does he have a British accent?

Anna: It's a very cultured, very trained voice. Very rich. It's like he belonged in the theater.

Budd: Let's go back to the image of him in England with the rugby shirt on. Where do you meet?

Anna: In the stands.

Budd: In the stands for a game or something?

Anna: Soccer game. I don't even like soccer! What are we doing here?

Budd: Does he tell you anything about himself in the soccer stands.

Anna: No-o-o-o-o.

Budd: Does he ask about you? Or were you just talking?

Anna: Yes, sort of.

Budd: How old are you?

Anna: Fourteen.

[Note: I actually knew I was eleven, but I lied, to make Brien think I was older.]

Budd: And he's maybe sixteen?

Anna: He's a lot older. I think he just thinks I'm a kid.

Budd: Okay, we're going to move back to the theater and he is there and he's told you he's going to see you again, and the little gray guys are going to take you away. Is that what happens?

Anna: God, that's a neat feeling! Rising above the steps. Ahhh, out the front door and up.

Budd: Now, when they take you away, where do you go? You're flying along, where do you end up at the end of this flight?

Anna: In the woods. It's all pine trees.

Budd: Are you still naked?

Anna: Yes.... Oh, my goodness, that's the pine tree where I buried the foal. It's in the back of the property.

Budd: Do you see Beth?

Anna: No. All of a sudden I got my nightgown back on.

Budd: While you're standing in the field?

Anna: Yes, it just sort of appeared.

Budd: What do you do next?

Anna: Walk back to the house.... Say "hi" to the horses on the way back. [*Surprised:*] Beth's in bed! She's already in bed. Shall I go back to bed?

Budd:	I want you just to look at yourself, your body, your hands, your feet, knees. Is there anything—
Anna:	My feet are dirty.... Oh, well, I'm tired. I'll go to bed anyway.
Budd:	Okay, just before I wake you up I want to set a little scene. A number of years ago, you were in a department store shopping. And when I count to three, you'll hear somebody calling to you very softly.
Anna:	*[Silent]*
Budd:	What are you hearing?
Anna:	Didi kitten.
Budd:	What is that again?
Anna:	Didi and kitten.
Budd:	What does that mean?
Anna:	*[Laughs]*
Budd:	What does Didi mean?
Anna:	I don't know, but there's somebody playing underneath the dresses over there. A little girl, in the store.
Budd:	Didi. Kitten. What's the little girl look like?
Anna:	She's blonde and she's got...saddle shoes on and socks.
Budd:	How old does she look like she is?
Anna:	Six.
Budd:	And she says, "Didi Kitten"? What do you think that means?
Anna:	Like she wants me to play.
Budd:	She wants you to play. Is she calling you like you're a kitten?
Anna:	*[Whispers:]* Yes.
Budd:	What's the Didi mean? Is that just childish words, or does that mean something to you?
Anna:	I don't know.... My mother takes me away too quick.

End Transcript

Budd brought me back to real time after implanting in me an awareness of my own health, that I could handle whatever these experiences were, and that I was a survivor.

After the session, Beth pointed out that "Dee-Dee Kitten" was the name of all of her stuffed toy kittens, and that she had saddle shoes and wore them all the time.

The similarities to Beth's recall indicated to me that we were remembering the same abduction in the large cavern. Beth also confirmed that the man wearing jockey shorts was also wearing slippers; she had held that back as her own means of testing veracity. We each saw only one man in jockey shorts; and depending upon which side we viewed him from (we saw him from opposite sides) his scar was visible either on the inside calf or outside thigh of his left leg.

Immediately after the hypnosis session we got out Budd's New York telephone books and looked for Brien Schultz. He wasn't there. I feel a kinship to him, whatever his real name might be. Have we been brought together since childhood as well? Okay, I'm hooked! I'm ready to meet him in this reality.

chapter 9

Denial is Alive and Well

47. Anna

After the hypnosis sessions with Budd, I thought things would get better. They didn't.

I found myself trying to accept my involvement and attempting to remember what had happened to me. The headaches eased up a bit for awhile, but then came back full force. Maybe Budd was right, I'm trying too hard. It didn't help that the gray shits kept up their intrusions. In December, they came four times in one week!

Bob Huff and I had been searching through my journal, looking for patterns. We found a disturbing one: For each memory that Beth had of an abduction in December and January, they came again exactly one year later! Of course, now that she was remembering more, and I was relying on my own intuition more, we found out that their visits were much more frequent than we had imagined. There was no way to tell if last year's visits were also more frequent than Beth remembered. For a while they came every Monday or Tuesday. When Bob caught on to that pattern, it stopped.

Every new bruise or cut on my body had to be explained. But too often I couldn't come up with a satisfactory, terrestrial interpretation. I couldn't explain the puncture wounds over the veins on my hands or in the crook of my elbow. I couldn't explain the strong vaginal discharge that smelled like semen, especially when I hadn't slept with a man in days and had taken two showers in the meantime. I didn't want to explain the scab in my navel that I found the same day. I couldn't explain the round itchy patches on the back of my hand, or the straight cut on my biceps that I noticed the same morning. At least I didn't have the severe things happen to me that Beth did. On different mornings, she had woken up with a ruptured eardrum, a hole in her calf all the way to the bone (she could barely walk for a few days), or with a red, bleeding rash on her arms.

I found myself trying to connect every odd happening to an alien abduction, yet also trying to deny that was the cause. It was all so confusing. I started having panic attacks. I'd be driving along in the car and feel the need to take a short cut or side road, or feel like I was being watched. It took an effort of willpower to stop myself from panicking. My imagination was running away with me. I was scared. I finally started yelling at the gray shits, in my mind, "Go away, leave me alone!" It helped, the panic attacks subsided. I will take control of my emotions, even if I can't have complete control of my body.

Beth's and my relationship also suffered during this time. We were both on edge and we began to squabble over almost everything. She had such a desire to explain it all away; I wanted to confirm that something was really happening to me. I already knew that her stuff was real. I wanted more memories; she wanted fewer. I turned to others for support. I told three other people, including my new supervisor at work, about the possibility of my involvement with alien creatures. I was relieved by their caring attitude and offers of support. In one case, my supervisor's attitude of "Oh, that's nice. Now, what were we talking about?", was upsetting, but at least he didn't reject me. My biggest fear was that once people knew about my strange other life, I would be more isolated from them. It was okay when I decided to isolate myself from others, but rejection was another matter.

It didn't help that I began to suspect that we were both frequently being taken on the same nights, maybe every time.

My work took me to Pennsylvania for two weeks and I thought that I would have time to sort things out alone. Nope. The greys decided against that. I was abducted three times in those two weeks and Beth was abducted from home the same three nights! So much for safety in separation. I seriously considered selling the farm and moving to a different location. Then a discussion with Bob brought a little reality into the situation. If this had been going on all my life, it didn't seem to matter where I lived, or where Beth lived, for that matter. When they wanted us together, they put us together.

There was no safe place for me.

48. Beth

What a surprise this must have been for Anna. Admittedly, I had thought Anna was involved, but I certainly didn't wish it on her! In truth, it was good to have a skeptical, detached friend who could play the role of investigator, as well as devil's advocate. Her denial had played well for my benefit, but it was selfishness on my part that allowed me to so appreciate her attentions! Anna's skepticism over her own possible abductions freed her to concentrate on my experiences, support me in times of overwhelming fear and withdrawal, and to always be there when I needed someone with whom I could speak openly. Even though I was confident of Anna's involvement (simply because of the many consistencies in our childhood memories and shared "dreams"), secretly—and perhaps unconsciously—I worried that the revelations at Budd's studio would alter that relationship, placing me in a more supportive role. I might not be able to offer her the same undivided attention. I was afraid I would fail her.

Getting back to a routine at the farm was proving difficult. The real world, which had been put on hold, rudely imposed itself like an uninvited guest at a private party. Why hadn't the world changed? Why hadn't these frightening and disorienting revelations of alien interference affected it? Everything was the same! The horses still needed to be fed and cared for, the stalls still had to be cleaned, the students still came for lessons. The sun came up, the sun went down. The world expected its inhabitants to go about their business as usual. But knowing this did not make it easier to cope. Logic had no place in these bizarre circumstances. Life would never be the same for us; our view of the world—though it had not visibly changed—would forever be slanted in favor of the unexplained. What we saw and what we did would always be questioned. Did we really see that? Did we really do that? Are we merely being paranoid, or are we really being watched?

As an independent-minded woman of the nineties, I resented the idea that I might not have full control over my actions and desires, that my environment may harbor entities who could pull my strings whenever they wished without either my consent or knowledge. At every step I found myself questioning my actions and reactions. Did that spot of blood on my arm mean something had happened the night before? Had I unknowingly put my nightshirt on inside-out before going to bed? Did I experience highway hypnosis or was I diverted by *them?*

As the weeks passed, though, I began to settle into my regular duties, making a deliberate effort to eclipse these uncertainties with the realities of day-to-day living. The cycle of life appeared to be unaffected by my personal trauma, and since I was certainly spending more time on this plane than the other, it was pointless to expend so much energy dwelling on it.

Besides, it was my turn to play devil's advocate, and I had to admit that the role was a challenge. Anna's emotions were in turmoil, her belief system in limbo. One moment she was fanatically obsessed with discovering "what *really* happened" behind every unusual event in her life, the next she would adamantly deny that anything unexplained *had* ever happened! I understood this confusion—I'd felt it myself often enough—but it was sometimes difficult to know when to play my new role, or even which new role I should be playing. Should I be the devil's advocate when she was in denial, or when she was in full acceptance? Should I instead just concentrate on supporting her, no matter which belief she currently embraced? I felt it was best not to openly display my own beliefs during these philosophical discussions, but Anna would often demand to know. I was caught between speaking the truth as I saw it or remaining uncommitted and detached.

This conflict provided me with an added insight: I could now empathize with the Fund's investigators in their struggle to maintain equilibrium in the face of so much emotional confusion. What I had initially seen as indifference (or perhaps reluctance to divulge information or venture an honest opinion) was now better understood as a sincere desire not to influence me—one way or the other. I had to decide for myself what was true and what was acceptable. And so did Anna. I could not decide for her.

I felt trapped between a rock and a hard place—a predicament Anna must have faced numerous times when dealing with my traumas. I couldn't continue to straddle the fence, but I was afraid to commit myself to either philosophy, finding that I had a propensity for taking the safest route. In self-defense, it seemed safest to deny the phenomenon, to classify it as common coincidence, or faulty human memory, or lucid dreams. Yet we both knew these explanations, as reassuring as they were, did not justify the whole. We had taken a chance just considering the *possibility* of alien abduction as an explanation. Until something better came along, something that could explain *all* the connecting events, we had no choice but to continue as we had been; examining each unexplained event, comparing notes on shared memories, talking candidly with family and friends, and keeping an open mind.

Expanding our horizons, however, did not mean we should accept every unexplained happening as being related to this phenomenon; we understood the value of skepticism and knew that a certain amount of it was required in order to maintain equilibrium. A healthy skepticism is the catalyst in the search for truth and understanding. Without it we would only be sheep following the herd.

49. Anna

As the months passed I began to get a few more glimpses of the memories that were buried in my subconscious. But they were never complete enough for me—I always wanted more information. When Budd Hopkins came to our area for a conference, I talked him into helping to retrieve some of those memories again.

We decided to go back to the suspected incident in December of 1993 when I felt that I might have been impregnated since my period was three weeks late in January. I felt the need to confirm what my intuition was telling me. I was so frightened when Budd regressed me back to that night in December that my body was shaking uncontrollably. Budd asked me to imagine that we were making a movie.

The hypnosis session was taped and later transcribed.

Saturday evening, February 6, 1993

Budd: Now there's a woman down there that looks a great deal like you, the actress that's portraying you. What's she doing as we look down?

Anna: She's lying there with her legs up.

Budd: Humm? Her knees are bent? Is that what you mean?

Anna: Yes.

Budd: Covers across her legs?

Anna: No covers.

Budd: Is she wearing a nightgown or pajamas?

Anna: Nothing.

Budd: Let's see if there is any movement. Maybe her knees move, maybe her hands move. Is there any movement there?

Anna: There's some bald heads next to her.

Budd: Bald heads next to her. Let's look down. Where are they?

Anna: One on each side, by her hips.

Budd: Do these bald things speak to her, does she have a conversation with them? Do you hear anything or sense anything?

Anna: There's another one by her head. It's taller.

Budd: These bald heads, do we sense any communication or is it just...?

Anna: No.... Bald heads. It's not fair that I have to do this again.

Budd: What is it that is going to happen now, let's watch and see....

Anna: There's another person coming in.

Budd: Uh-huh.

Anna: Two gray guys, one on each side.

Budd: Is that Beth coming in?

Anna: No, it's a man.

Budd: Okay. What does this man seem to look like?

Anna: He's got red hair.

Budd: And what kind of clothes does he have on?

Anna: There aren't any.

Budd: When he comes in, does he speak to the woman on the bed?

Anna: No.

Budd: Does he know who she is? Are they friends?

Anna: Yes.

Budd: Yes, okay. Let's just look down and you can see.

Anna: They told him he had to be here, too.

Budd: Who told him he had to be here?

Anna: The gray guys.

Budd: Is he happy to be there?

Anna: No.

Budd: Let's look at the man. Does he come around and chat? Or does he stand?

Anna: He stands at the end of the table, between her legs.

Budd: Uh-huh. Now very coolly, you have a very clear mind. Let's see if we can read any expression on that man's face.

Anna: *[Crying:]* I don't think he wants to be here.

Budd: Um, what, is there something that makes you feel that as you look down on him?

Anna: I can't decide if he's just sort of like a zombie, or he's trying to fight them, but he can't.

Budd: Does he turn and say something to the people on either side of him?

Anna: No. Pushed him.

Budd: What? I'm sorry?

Anna:	It's over.
Budd:	What?
Anna:	They pushed him forward.
Budd:	They pushed him forward. Okay. And then what happened when they did that?
Anna:	His penis went in her. And they took him away.
Budd:	Uh-huh.
Anna:	They had to push him.
Budd:	So he had an erection?
Anna:	Uh-huh.
Budd:	When he is pulled out, is there any sense that he either ejaculated, or didn't ejaculate? Maybe we can't tell.
Anna:	I think he probably did.
Budd:	Okay. And after they, after he was out, what did he do?
Anna:	They just backed him up out the door.
Budd:	They back him up. Did he walk backwards?
Anna:	No, he doesn't walk, he just goes backwards. He's not real happy, either.
Budd:	Let's look at the woman on the bed, the one that looks like you, playing you. What is her reaction?
Anna:	She's glad it's over.
Budd:	This is very important. The woman on the bed, the table—whatever—perhaps can't really say much. But, let's right now imagine what she might want to say to the man if she could speak. What might she say to the man right now?
Anna:	*[Crying:]* I'm sorry. I don't want to be here. I'm sorry.
Budd:	Do you think it's her fault in some way that he was there?
Anna:	Yes.
Budd:	Did she ask for him to be there?
Anna:	No.
Budd:	Then it's not her fault. She feels very sorry, doesn't she? But she has no reason to feel sorry, it's not her fault. It's not her fault that she is there. It's not her fault that he's there. I want the woman on the table to speak to the men up there, the men that brought her in and brought the man in. What would she like to say to them that she couldn't say then?

Anna: Leave us alone! Don't come back. Don't do this again. *[Sobbing:]* Leave us alone!

Budd: Watch this scene, now it's ending. How does it end, exactly?

Anna: She gets up off the table.

Budd: Uh-huh. She does.

Anna: They put her nightgown back on her.

Budd: How does her body feel?

Anna: Sore.

Budd: Sore. Where does it feel sore?

Anna: Here. *[Places hand on stomach]*

Budd: In her stomach?

Anna: Yeah.

Budd: What made it sore?

Anna: Shit!

Budd: What do you think would make it sore?

Anna: The tube again.

Budd: What was the tube? I don't quite understand.

Anna: Looks like a real thin 20 gauge needle, but it's about two feet long.

Budd: What was on the other end of the tube? Can we see?

Anna: Yeah. It's where it goes behind me.

Budd: I see. Okay. They do this before they brought in the man or after?

Anna: No. Before. Oh!

Budd: Does it hurt?

Anna: It hurt!

Budd: What's happening?

Anna: Cramps.

Budd: Cramps? You're having them? What are those cramps like?

Anna: Just mild.

Budd: Just mild. Are they like menstrual cramps or in a different area? Intestinal?

Anna: No. Like menstrual cramps.

Budd: Menstrual cramps? Do they follow the tube going in? Is that afterwards or before?

Anna: It's after.

Budd: Now she's standing up now and you're looking down. She's got her nightgown on. She's feeling a little better now. Let's see what happens next. Every dream or memory has a sequence, so she's standing here, let's see what happens next.

Anna: Hum...they float her away.

Budd: When they float her away, what does she see? You can see as she's moving along.

Anna: Looks like people in beds in the hallway, in—in the wall! (See Figure 10.)

Budd: Uh-huh.

Anna: Like they're waiting or something.

Budd: Do they have pillows and blankets?

Anna: No! They're not very economical, I mean, they've got them sideways. If they were going to be in the wall, you'd think they'd put them in feet first or something.

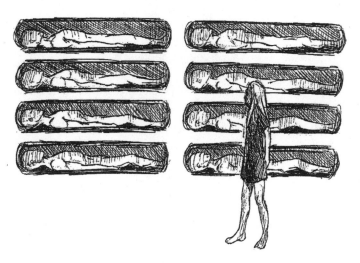

Figure 10. Alien storage chambers witnessed by Anna during an abduction.

Budd: As we go by them, let's look down and see if we see any-body we recognize.

Anna: These people are bald! This is weird.

Budd: Are all of them bald?

Anna: Yeah!

Budd: What are they wearing?

Anna: All in gray. They're not people!

Budd: Are they just resting there? Sleeping?

Anna: They're not moving, they're just there.

Budd: Just there. Get a good look at how the beds are arranged.

Anna: It's like holes in the hallway.

Budd: Okay. Let's see where she floats to now as she moves along.

Anna: That room again with the view screen.

Budd: Uh-huh.

Anna: *[Petulantly:]* "No, I still don't know any more about stars. But I don't want to know either. Why don't we do something else?"

Budd: What do they do about the stars?

Anna: Jo thinks I should know more about the stars. Told him I don't care.

Budd: Who's Jo?

Anna: He's always there.

Budd: What does Jo look like?

Anna: He's a big grey.

Budd: You say there's a screen, a view screen. What's on the view screen?

Anna: Just blips of light. Like stars. Constellations. But I haven't seen those before.

Budd: Uh-huh.

[I moved on from that location to another spontaneously.]

Anna: Oh, gold dust all over. Neat!

Budd: Gold dust, did you say?

Anna: *[Wonderingly:]* Yeah. It's almost like it's liquid, but it's in the air. Comes up from the bottom, sort of gold flecks.

Budd: Is it like snow particles?

Anna: Yeah, it's like snow coming from the bottom up. Tastes funny. "You didn't tell me I wasn't supposed to eat it!" He won't tell me what it is. It's supposed to be some sort of cleansing. *[Rubs eyes]*

Budd: What's happening to your eyes?

Anna: They're sore again.

Budd: Is it from the gold dust? Or you don't know?

Anna: I don't know, it could be.

Budd: Is there anybody else in the room with you and Jo, or is it just the two of you?

Anna: I don't know, the room is so big I, it's like I can't see the other side. So big. Sort of like a catwalk, we're moving around. Yes, there are people on the other side. There's a littler ship inside, in the middle. I think I'm supposed to go down to it.

Budd: Is this ship about the size of a little compact car?

Anna: No, it's like a big station wagon, 20 to 30 feet. It's spinning and there's lights.

Budd: Uh-huh.

Anna: He doesn't want to jump over the edge. That would be fun. Guess we'll have to go hike down to it.

[At this point I become very agitated and Budd attempts to calm me.]

Budd: Just relax, take a very easy, deep breath now. Just tell what is happening now, Anna.

Anna: Oh, the children again.

Budd: The children. Just tell me what you're seeing.

Anna: Oh, I'm in that room again with them. *[Starts to cry:]* They're all over the ground!

Budd: Anna, feel my hand on your hand. You're here with me. Just close your eyes, just relax. Just tell what you see in the room. We're just looking back. This happened weeks and weeks and weeks ago. This is the present now, we're just looking back. What's this about the children now? I don't understand.

Anna: Oh, it's just that room where they all are. Oh. The little ones that walk all around and grab hold of your legs. *[Crying]*

Budd: How does that make you feel?

Anna: Terrible.

Budd: If you had your wish, what would you do right now?

Anna: Never to see them again. There's older ones here this time. Oh. They're taller.

Budd: Are they smiling and happy to see you?

Anna: No.

Budd: Children sometimes laugh and play. What do they do?

Anna: Nothing. 'Cept, just grab a hold of you.

Budd: Why do you think they grab hold of you?

Anna: Oh! They want somebody to love.

Budd: Do you feel you can give them love?

Anna: No.

Budd: Why don't you think so?

Anna: I don't like them. They scare me. Oh!

Budd: Do you communicate with the children? Do they communicate with you in some way?

Anna: Somebody says that they belong to me. I don't believe that.

Budd: There's no reason for you to believe it. You believe exactly what you want to.

Anna: I don't want anything to do with them.

Budd: What I want you to do for me right now, Anna, is to pick out one in the group and look at one. Look at one closely and describe that child to me. If it's a boy or girl, or age, or size. Just pick out one.

I described a girl about six years old. The girl had very thin, almost white hair, large, very light blue eyes, no eyebrows or eyelashes, and very small ears and a little mouth. Overall, her coloring seemed washed out.

Budd: Does she have a name?

Anna: The name that came to my mind was Margaret, but that's my mother's name. *[Crying:]* She can't have that name!

Budd: Why do you think you thought that name?

Anna: I don't know! I think that's her name. It can't be.

Budd: Did she tell you that, or did someone else tell you that?

Anna: I don't know.

[My mother died about six years before. I believed I was told this child was mine and that I had named her Margaret.]

Budd: What's this little girl wearing?

Anna: Oh, white thing. Sort of loose.

Budd: Does it have a belt and pockets?

Anna: No, nothing. Like her mother dressed her in a sheet. Except it's real thin.

Budd: Let's look at the face again. I know it's hard for you, but you can look at a face and judge the expression. What's on her face?

Anna: Confused.

Budd: The face seems confused? What do you think she thinks about you?

Anna: She doesn't know who I am. *[Crying:]* "I don't look funny!"

Budd: I'm sorry?

Anna: She thinks I look funny.

Budd: How can you tell that?

Anna: I think she told me.

Budd: Uh-huh.

Anna: I'm so big and ugly. "I'm not *big and ugly!*"

Budd: How many children are in this room that seem to be feeling some sort of connection with you, do you think?

Anna: Eight or ten.

Budd: What does their skin feel like when they touch? Does it feel warm?

Anna: No, it's like dried out leather. It needs some oil.

Budd: Very dry skin.

Anna: Yeah. They look like they should be smooth, but they're not.

Budd: How about their hands? Let's look at their hands.

Anna: They don't seem to be pudgy. Little kids hands are supposed to be pudgy, these are not.

Budd: Are they the same kind of fingers like you have and I have?

Anna: I don't think so. I mean, they're different somehow. I don't know. They're backing away.

Budd: They're backing away?

Anna: Something's hurting them. Oh. The taller one that sort of looks like them. They told me that wasn't such a bad ordeal. I said, "Yes it was. Don't do that again to me."

Budd: After they move you out of the room, I mean after they move the children out, what happens next?

Anna: Down another tunnel, but this one is red.

Budd: Uh-huh.

Anna: And there's blue something or other.

Budd: What just happened?

Anna: I went through the window.

Budd: You're at the window?

Anna: Through it.

Budd: Through it. You're at your bedroom?

Anna: Uh-huh.

Budd: Did you feel it?

Anna: Yes.

Budd: What did it feel like?

Anna: Kinda tingly. It feels kinda good, like itching on the inside. You get to scratch everything.

Budd: Where are you now?

Anna: Standing beside my bed.

Budd: Let's look at the bed, see how the covers are. Are they all pulled back on the floor?

Anna: No, they're all sort of bunched up.

Budd: What's happening now?

Anna: I'm trying to figure out how the covers get all messed up. I don't know how that happens. It's like somebody had a fight in there.

Budd: Uh-huh.

Anna: I guess it doesn't matter.

Budd: Maybe you should just get in bed. Do you want to get in bed? Are you in bed now?

Anna: Okay.

Budd: Okay. Just before you drift off to sleep, this is what I want to do. I want to go through your body and see how your body feels.

[Everything felt normal, except for some abdominal cramping.]

Budd: Anything feel different about your head?

Anna: My nose hurts.

Budd: Let's do the features. First, your eyes. Anything at your eyes feel different?

Anna: A little scratchy.

Budd: A little scratchy, okay. Now let's go to your nose. We'll do this very systematically. Inside and out. You said your nose hurts.

Anna: It hurts inside.

Budd: Inside. What do you think caused the hurt?

Anna: They must have done something in there.

Budd: Let's move to your mouth. Anything different at your mouth?

Anna: It still tastes funny.

Budd: What does it taste like?

Anna: *[Smacks lips]* Almost like a salty alum. Not quite that bad. Not that dry.

Budd: Okay. Let's move on to your ears now. Anything different about your ears, inside and out, or do they just feel normal?

Anna: Ringing.

Budd: Okay. I want all of those sensations just to ease off. I want you to wake up with a sense of relief—a feeling that those mysteries and problems that you remembered partially back in December, now you know what happened. I want you to wake up with a great sense of relief that you have gone through this great, trying, complex experience and you have survived again.

End Transcript

Afterward we discussed the things I had recalled under hypnosis:

Anna: I guess my suspicions were right. I wish they'd quit doing those children things. Aren't I too old for that?

Budd: I sometimes think they have a whole backlog of ova locked up.

Anna: They've got to. They don't need us.

Budd: It's almost like they don't know the optimum way to do this. It's like they see if they can get the human to hug the child, the child will live and be okay. Like magic. Where if they bring in a man and have an act of two-second intercourse that somehow that's going to make things more—

Anna: Whatever.

Budd: Make the pregnancy take better or something like that. Tough. So, he has red hair, huh?

Anna: Yep.

Budd: Have you ever seen him before?

Anna: In real life?

Budd: Uh-huh.

Anna: No, just with them. It was Brien.

Budd: Want to go into the other room? Can you walk?

Anna: Yeah, I'm okay.

End Transcript

I really hate what the gray ones do to my mind, even more so than what they do to my body. The continual episodes with the children terrify and exhaust me. Having me name the six year-old after my mother is pure psychological harassment. The continued meetings with Brien, and the sexual relations (I feel strongly that he is the father of all my children that they have produced and kept), are humiliating. They are not enjoyable. I never thought it was possible for two people to be raped in the same act; now I know it is.

50. Anna

The hypnosis sessions with Budd, self-hypnosis, and spending time relaxing my body and mind began to pay off. I found myself less angry with the gray ones, less angry with Beth and myself, and remembering more.

One afternoon as I was meditating about the possibility of Beth and I knowing each other in childhood, I came up with an interesting memory. I recalled a gray wall with a stick protruding from it. Beth was there. I was about three years old, she was about six. The stick was maybe an inch in diameter, maybe yellow, and about a foot long. I saw it rotate in the wall. I made it move, even though there was no indication that it could move within the wall. I could see myself going up to it (levitating] and hanging there while Beth stood beneath to catch me when I dropped off of the stick. (See Figure 11.)

When I told Beth about it later that afternoon, she told me what she remembered during the hypnosis session with Budd. I had been asleep during her session, and we had not talked about what had been revealed. Beth remembered that we were supposed to manipulate the stick with our minds; make the tip of it inscribe large and small circles, go fast and slow down. We were then to make it move down the wall, grab it, then levitate it and ourselves back up the wall. Close enough to my memory for me to believe it really happened.

Figure 11. Anna playing with an alien "toy."

Beth then asked me to think of a box, and then do something with it. I immediately said, "Turn it inside out with my mind." She said that's what we were supposed to do with it. Beth also asked me if it changed when it was inside out. I said, "It gets smaller, but that's impossible." Beth said no, it did get smaller. She asked me what color the box was. "Gray," I told her, but she hadn't been able to remember the color.

Beth asked if I remembered a room and someplace in it that we weren't supposed to go. I "saw" a foggy area in the center of the room. She said no, not in the center, someplace else. So I looked around the room and saw a mirror, but was amazed when we walked through it! Beth confirmed that, although she remembered us entering the side of the mirror instead of walking straight through it. She asked me what was on the other side of the mirror: "A corridor, with viewscreens on the walls." What was on the screens? Lots of things—meadows on one, animals on another, fire/orange stuff, birds, mountains. The scenes changed.

I remembered myself running up and down this corridor, pushing plates beneath the screens and Beth yelling at me to stop doing that and come back, which I eventually did. Beth's memory of the situation was of a "blue-light corridor." I just remembered it as being dark, but somehow filled with light. She said we were supposed to change the pictures with our minds; we weren't supposed to push the plates under the screens. If we walked by the screens and thought of something, it appeared on the screen. Yes, she remembered yelling at me to stop running up and down pushing the plates.

Beth then asked me if I remembered any other "toys" we had been given to play with. I remembered a giant "hamster wheel." I sat in the middle of it and rolled it by climbing on the rungs. But I had much more fun sitting in the middle of it and levitating up and down through space! Beth told me, "No, what we were supposed to do with it was to sit in the center of it and make it move around us without touching it."

I don't like doing mundane things. I push things to their limit. I'm always looking for new uses for ideas and objects. I had to find more fun things to do with their toys. Beth said she was always yelling at me to do things right! She was so afraid I would get us in trouble with the aliens. Maybe I did; I don't remember.

I also remembered the "fire stick." That was really scary. We were put in the center of a room with a grey standing off to one side. A fire stick was whizzing around the room. Beth and I had to concentrate, together, to keep the stick from touching us. It took both of our minds to control it, to throw it in another direction without touching it. It hurt when it touched us. When we threw it, it went in another direction than what seemed to be normal. We'd throw it in front of us, but it would go behind us or off to the side. It was so hard. We stopped the game quickly when we learned how to aim it at the alien. He caught it and we could relax.

Beth then decided to write something down, "to make it more scientific," before I answered. She then asked me if I remembered going out of that room to somewhere else. I went up a ramp (catwalk), down a side corridor, through a laboratory area; an operating room with instruments on the walls and ceiling (some were lights), and tables against the wall. Then I went into the "control room" where the star map viewscreen is; then into a nursery with the kids; and then into a place where they had lots of animals—they weren't in cages, but they weren't free either. I had the image of one of the greys taking me back to the room with Beth, telling me that I wasn't supposed to be roaming around by myself.

Unfortunately, our scientific experiment proved inaccurate; this wasn't what Beth had written down. She had walked down a ramp (floated, actually, throughout all these scenes) to a dead-end where there was a large black triangle on the wall with patterns of white lights all over it. Oh, well.

I felt elated after realizing that my memories were validated by Beth. If she had not been there, I would have thought they weren't real memories, just other strangenesses of my mind. If I was alone with this phenomenon, it would be easy to overlook memories like this, and just think my mind was making things up.

I checked out the stick in the wall and the box by performing a random survey of ordinary people over the next few weeks. They all came up with normal things to do with them—cover the stick, break it off, hang from it; squash the box (flatten it), throw it, or open it up. About a year later, I talked to a man at a Fortean conference who was excited about the stick in the wall, the box and the fire stick. He said that mathematicians have predicted objects could be manipulated in exactly these ways—in the fourth dimension (or hyperspace). He even gave it a name, but I have since forgotten what it is.

On another afternoon, Beth and I were talking about childhood memories and we both remembered going across a field with another little girl. We each, independently, wrote down what we were wearing and what our companion was wearing. They matched. A few weeks later, Beth brought home some pictures of herself that her parents had kept. Among them was a picture of the little girl that I had seen that day, wearing the same dress and shoes! It was Beth. Very spooky.

I have now decided that I may have lots of childhood memories. The more I search, the more I have. Many of them quite pleasant, having nothing to do with alien abductions. It's a good feeling. I'd just never thought much about my childhood before. I'm not one to dwell on the past; I live for now and the future.

Yes, Beth and I have been brought together throughout our lives. We have known each other forever. But why? What is the overall plan that required the gray ones to go to so much trouble to bring us together, to educate our minds, or to take such risks that we would be missed? Was our recent memory recall intentional or has it upset their plans? Is it all part of a larger design? I wish I knew.

51. Beth

Christmas was only a few weeks off, but I didn't feel at all jolly. I was still experiencing problems digesting meat products other than beef, no matter how cleverly disguised. When Anna and I attended a banquet a few weeks after returning from our visit with Budd Hopkins in New York, we were offered two main dishes: meat loaf or chicken; I opted for the meat loaf, confident in my choice, but barely had time to finish eating before I was rushing to the bathroom to get rid of it. Later, we discovered the dish included pork and veal as well as beef!

As long as I stayed away from other meats (in any form) I had no problems with my diet or digestion. Budd's lengthy post-hypnotic suggestions that I would be able to eat whatever I liked seemed to have fallen on a deaf stomach. Either that or the aliens had discovered a much more effective means of implanting a very powerful suggestion. Not being able to comfortably digest non-beef products caused me to crave the stuff at all hours of the day and night! I was beginning to dream about eating fried chicken, roast turkey, pork tenderloin, veal chops, hot dogs and even chicken livers (which I had previously detested)!

To add insult to injury, the weird visitations continued unabated. At 4:00 A.M. on December 5, 1992, I awoke to find myself sitting cross-legged in bed with my nightshirt turned inside-out. I had a funny, acrid taste in my mouth, and my stomach was queasy. I had no idea what might have happened, no memory of the little gray shits invading my sleep, but within a few days I recalled fragments of what might have taken place on that night: I remembered, through flashbacks, standing near Anna and watching as one of the greys injected something into her right thigh. It left a prominent puncture mark, though Anna did not seem to be aware of the intrusion at the time. When I told Anna about my flashback, she examined her thigh and found a small, red-rimmed mark which appeared to be a puncture. It was exactly where I recalled seeing the injection given.

I also recalled being forced to drink a vile tasting thick liquid which suspiciously resembled molten lead, and probably tasted worse. I knew I had been made to drink something similar during past abductions and wondered what purpose the drink served. It invariably made me nauseous, and the foul taste stayed in my mouth for hours afterward. If their intent was to make me sick, they were batting a thousand.

The onset of the holidays contributed to a mounting stress that was becoming unbearable. I had withdrawn from the world, to all intents and purposes, unable to cope with the mounting tensions. Hypnotically retrieved memories, flashbacks, Anna's private torment, the workaday world's oblivion, the upcoming Christmas holidays—all of these joined forces to remind me that I wasn't as strong as I thought. In retaliation, I withdrew into myself.

As my already fragile emotional state deteriorated, friends and family began noticing the change, yet I remained unconcerned. Certainly I was aware that the stresses were building, but they still seemed no threat to the fortress I had erected around my psyche. I was a pragmatist, a realist. If these flashbacks were real memories of real events, then I would learn to deal with them in time. But we had no proof of that, so both Anna and I were still somewhat detached from it.

When anyone commented on my apparent absentmindedness, I would calmly announce that I was just tired—or, to those aware of our situation, that I was working on compartmentalizing these experiences and that took a great deal of concentration and resolve. To get my act together, I would need time and space. I needed to be left alone.

One afternoon I found myself curled up on the floor in the corner of my bedroom. I don't know how long I might have been there, squatting in that fetal position, but when realization broke through I knew it was time to get help. The fortress had fallen in the attack and my mind was in imminent danger. It was hard to accept that I had slipped so far so fast without my being consciously aware. Subconsciously I had obviously taken pains to regress in private, somehow knowing that this collapse would place Anna in the position of having to support me once again. What help was I to her when I couldn't even help myself?

A call was immediately placed to Dick Hall in hopes of obtaining the name of an area psychologist or psychiatrist who could help me. He obliged me by suggesting I contact Dr. David Ruxer, a clinical psychologist about an hour's drive from us. He was expensive, we were warned, but had worked with other abductees in the past. It concerned me somewhat that Dr. Ruxer might be a "true believer," therefore more concerned with the content of my experiences than with therapy. This, Dick assured me, was not the case. He had, in fact, not accepted any new patients who reported such incidents in the past several months because, as Ruxer himself admitted, he was having difficulty dealing with the emotional backlash from these sessions. Would he even see me, then? And if he did, could he help me?

I didn't need a therapist to confirm or deny the existence of aliens. I didn't care at that point whether my memories were real or imagined. I only wanted help learning to cope. I wanted to be the way I was!

On December 9, 1992, I entered Dr. Ruxer's office, having successfully fought off an overwhelming desire to run away. David Ruxer was much younger than I had expected. With a warm smile and easy disposition, Dr. Ruxer quickly endeared himself to me, and I realized I had made the proper decision in coming to see him. Even if he couldn't help me in the long run, I felt better for having met him.

We spent several minutes just getting to know one another. In short order I was comfortable enough with him to admit how vulnerable I was, how close I had come to complete withdrawal, and how important it was for me to be in control of my emotions once again. I insisted he understand my priorities in coming to him for help: I did not need validation of my experiences, nor was I particularly concerned whether or not he believed in alien abductions. The only close encounter I was interested in was the one I had with emotional breakdown. That was too close for comfort and an obvious early warning that my state of mind was in jeopardy.

Ruxer's reaction to my monologue was intriguing. He frowned slightly, then nodded, apparently in agreement. There followed a moment of silence, while I worried that this might have been a mistake after all. Finally, he leaned forward in his chair, cleared his throat, looked me square in the eye, and smiled disarmingly.

"Thank goodness!" he sighed. "I was afraid you were going to ask me to regress you!"

Dr. Ruxer was a qualified hypnotherapist, but his policy was not to use hypnosis in the initial session, since it may prove unnecessary in getting to the root of a problem. He much preferred to help his patients develop effective coping mechanisms for keeping in touch with their emotions. He admitted that this was treating the symptoms rather than the disease, but had found that more often than not, when the patient learned to deal with the emotional responses, the core problem—the "disease" itself—was revealed. Once out in the open and recognized, the patient had the means for healing himself.

This, I thought, was a novel idea. I liked it. It seemed plausible that my feelings of helplessness might stem from losing touch with my true emotions. I had been trying so hard to remain detached that I had removed myself from myself. In response, my mind went along for the ride.

Relieved that I didn't expect him to confirm or deny the phenomenon, validate or disclaim my memories, Ruxer suggested a number of ways in which I might reconnect with my feelings. One method, which proved extremely helpful, was to ask myself these questions every waking hour on the hour: How am I feeling at this moment? What has happened that has caused me to feel this way? Would I prefer to feel differently? If so, how would I prefer to feel? Often, he went on to explain, when an emotion is put into words and its reason for being understood, it can be either enhanced or altered, just by verbalizing it.

"Have you ever felt uncomfortable around a person who always seems to be depressed, who never smiles or has anything good to say about anything?" he asked.

I admitted that I had. "I make it a point not to spend much time around those kinds of people. They drag me down with them."

"Exactly! You're responding with negative emotions even though you may have no idea why this person should affect you this way. You could be having the best day of your life, but you'll walk away from this encounter feeling depressed and unhappy. The emotions are real, but they didn't originate with you; they were a response to someone else's feelings."

This analogy was carried one step further: "Suppose this depressed, unhappy person was you, but you didn't want the person you were talking to to know you were miserable. How would you handle it?"

"I'd fake it!" I replied quickly, getting into the game.

"Right. And how would you do that?"

I thought about it a minute, visualizing this encounter. "I guess I'd act it out. Smile a lot, maybe laugh out loud. You know, act like everything was hunky-dorey, life was great. Is that what you mean?"

"That's what I mean. Now, if you kept this up, say, for several minutes, maybe even longer, would it alter anything about how you felt initially?"

"I suppose it would have to, after a while. I couldn't keep smiling and laughing for very long without it affecting me inside." Suddenly the light bulb flashed on. "I see what you're getting at."

"Good," he said, grinning broadly. "And that's what I want you to try to do, starting right now. But," he warned, "keep in mind that this technique is not an excuse to ignore why you're feeling depressed or withdrawn; it should be used to recognize what you're feeling and why you're feeling that way."

I knew what he meant. It was easier to handle emotions when I was not actually experiencing them! Looking back on what I was feeling—and why—could help me to get these feelings of helplessness under control and take the necessary steps to learn to cope with them.

He asked me how I was feeling at that moment. I felt pretty good, relieved. Why did I feel that way? Because I had something to work on, a way to help myself, to take control of my life again. Under the circumstances, it was not necessary to ask myself the other questions. I would have many opportunities in the future to go through the whole list!

I left his office feeling remarkably refreshed and anxiety free. I was not naive enough to believe that my problems were over after one hour in Ruxer's office and a few creative coping techniques. The problem—the disease—was still there and may never be cured or even understood, but at least I had a weapon to use against the symptoms now, and I planned on following the doctor's orders to the letter.

52. Anna

As the months dragged on, I kept getting confused. Despite my returning memories, I still had the feeling I was making all this up. I had a need to deny my involvement, yet I couldn't write all these things off to coincidence. That was too much for my logical side to accept. But I kept looking for another explanation.

There was stress. That could account for some of the physical manifestations: the headaches, diarrhea and exhaustion. Maybe I was a hypochondriac; although several doctors I asked said, "No." But neither explanation accounts for the bruises, cuts and scars.

Maybe I was into self-mutilation, yet didn't know it. But no one had ever seen me do anything like that to my body, and I had no memory of it, nor any desire to do so. I like my body, most days. But that didn't account for the strange memories.

Maybe I had a fantasy prone personality. I had read they exist, and firmly believe that many good fiction writers have this tendency. Maybe I couldn't distinguish between fact and fiction, maybe I had created an imaginary world into which I could retreat. I would prefer to think that I might fantasize about a more pleasant place where I was the queen, everyone and everything was beautiful, and instant gratification of my needs was the norm. My fantasy prone personality must be really sick to create rape, mutilations, ugly creatures and stark surroundings. But I wasn't particularly imaginative; I was pretty well grounded in reality—until a year ago. But fantasy doesn't account for missing time.

Multiple personalities could account for the missing time. In many cases, one personality blanks out when another takes over. But why would a second personality only take over in the dead of night, or only when I was alone? Why hadn't my co-workers seen some indication that something was terribly wrong? But a therapist, specializing in multiple personalities, had already eliminated that possibility for me.

Maybe I had a brain disorder, like temporal lobe epilepsy. I researched how the brain functions, and was amazed at how much we don't know about it. The temporal lobe is the area of the brain where sense is made of perceptions. People with temporal lobe dysfunction report hallucinations, déjà vu, panic attacks and even psychic or religious experiences. This could explain some of Beth's or my experiences, but for both of us? Many of our friends? Was temporal lobe epilepsy catching, like the flu? Maybe we have a brain disorder they haven't catalogued yet. But no brain disorder could account for terrorized dogs and horses, or the mechanical and electrical malfunctions of vehicles.

Maybe the memories were just hypnogogic or hypnopompic images as some skeptics claim. These seemingly very real images occur just as one is falling asleep or waking up. They happen to a lot of people. I have them, but they feel different than these other memories. Besides, hypnogogic images haven't been reported to last for hours in wide-awake people in the middle of the afternoon.

Maybe I was craving attention and had a deep-seated desire to be noticed, to be important. I think there may be less stressful ways of doing that, ways that don't require so much emotional trauma or the collusion of so many other people. Besides, I was well respected in both my work and my horse business, and my basic personality (introverted) can only stand the spotlight for brief periods.

Maybe it was all a product of too much reading on the subject and leading by the hypnotists. But Beth's *conscious* memories were what got us into this. Neither of us had read about abductions before last year, and she still hasn't. My involvement could have been triggered by my reading and my closeness to her, but I keep coming up with memories about things I'd never read about. When Budd asked me to draw the "squiggles" I'd seen under hypnosis and then showed me the same designs drawn by three other abductees, I was hard pressed to believe he had led me to create an abduction experience. Our conscious memories of the same experiences when we were children are also hard to explain in terms of what I logically know.

I decided to try another therapist, ostensibly to lessen the headaches and deal with digestive problems. I'd given myself ulcers by taking too much aspirin and the Zantac was working on them, but Tylenol wasn't doing as good a job keeping the headaches at bay. This was to be a very different type of therapy than I'd heard about, but I was willing to try almost anything. The therapy involved creative visualization and hypnosis.

The therapist asked me to go inside myself and concentrate on finding what was controlling the headaches. She explained to me that the subconscious knows what is happening in the body and I could visualize what was happening. This was definitely leading, and I wanted to follow her instructions. Once I found the "headache manager," I opened my eyes and described it to her. I visualized a large black shaft, like a tree trunk, moving swiftly up out of pinkish-red ground.

She then told me to ask it if it was willing to communicate with me. When I closed my eyes and concentrated, I saw a bunch of flying saucer lights spinning like mad around the top of the shaft! She then asked me to go back and ask if this was a "yes" response. Okay. Then we tried to find out what a "no" response was: The shaft started to sink back into the ground, and it also had a brown scraggly mushroom-like cap on the top. She then asked me to ask it what its "positive intent" was in providing the headaches. I checked: Protecting me. Protecting me from what? The image I saw was of a *giant* volcano spouting debris and smoke, with a red sky all around it! That shook me up. I didn't realize my memories were that frightening, or that volatile! I cried.

The therapist then told me that there were other entities (parts) within my body that could help the headache manager to come up with other ways of protecting me, ways that didn't involve head-aches. She suggested that I have a creative part, and also an historian, that knows everything I've ever known or felt, and that there were others that could also help. So I closed my eyes and tried to see all these entities together. It was a wonderful party. I imagined the historian as a crusty gray haired wizard with a large book, and the creative side was a small pinkish-red-yellow-purple-opalescent rose bud at the base of the shaft.

When I opened my eyes and told her about the party, she had me close my eyes and ask them for some practical solutions to the headaches! That didn't work; all they did was party. At the end of the session she told me to ask the entities to work on this problem without me—in the subconscious—that I had to leave now, but would come back and visit and find out what they'd come up with. I found leaving them to be very sad—it's like in that short hour I had found some very dear friends! I cried again.

After a cup of tea, we tried again, but this time we concentrated on whatever was causing the digestive upsets. I couldn't get an image of anything, but I felt very strong waves of dizziness. I'm glad I wasn't standing up! The therapist had me close my eyes, thank the entity for communicating with us, and ask if the dizziness was a "yes" or a "no." All I got was another very powerful dizzy spell. I guess that was a "yes."

I felt that this entity was like a little kid just so excited about being able to communicate with me that it did everything in too expansive a manner. Asking what a "no" was, all I got was complete silence, no imagery, nothing. I asked the entity for its positive intent in giving me cramps and diarrhea. Protection again, but I couldn't figure out from what. I was getting tired so we decided to quit. I closed my eyes to say goodbye, and got a very clear image of a teeny, tiny green-scaled fire-breathing dragon with a red tongue curled up in a tiny cave. Amazing what the mind can do!

In the weeks that followed, the headaches became less frequent—I only needed to take 1-2,000 mg. of Tylenol a day. The new way the dragon protects me is to give me cramps before I eat, sometimes. That seems to happen only when I want to eat meat. I am getting closer to a vegetarian diet than I ever have been in my life! I'm not sure I will like this. I can eat all the carbohydrates and sugars I want; meat seems to be the only thing I'm warned away from. I haven't had near as much diarrhea, but sometimes the cramps are still severe when I eat.

I talked with the therapist about what we had done and whether or not it was hypnosis. She said I was definitely hypnotized when I did the sessions with her—she could tell by the pallor of my face, the relaxation and my breathing. At times she said I was in quite a deep trance. All I ever thought I did was to close my eyes, relax and look into my mind!

There was a real difference in the hypnosis sessions I did with this therapist and the ones with Budd. I knew I was making up the stuff about the volcano and the green dragon. It wasn't real; it was just imagination. The sessions with Budd seemed so real, the emotions so strong; the remembered pain in my body was genuine.

So, I remain confused. If I could accept all of these other explanations, I could easily deny that I was being interfered with by alien creatures. But I can't; I've tried.

chapter 10

Reluctant Acceptance

53. Beth

My son Paul is a very serious-minded young man, the kind of individual who laughs infrequently because to him life is a serious matter. When he was growing up, I had worried a lot about his attitude, his lack of humor. As an only child, and growing up without a father figure, Paul had been forced to take on responsibilities way beyond his years. Although he handled these responsibilities well, showing a maturity that made me proud, there was a price to pay—for both of us. Paul missed out on much of his childhood, through no fault of his own, and although he seldom complained, as he matured into adulthood he must have realized that he would never get this time back. But he *could* make sure his daughter didn't miss her childhood—or the attention and guidance of her father.

It had taken me over a year to get up the courage to tell my son about this strangeness. I would have preferred never to tell him at all, but I didn't want him to resent me later, the way I had resented my father. It would have been easy to fall back on the same excuses: What was the point of revealing all this when Paul may never remember anything? Why should I take the chance of destroying our relationship—not to mention my credibility—when it may prove to be nothing more than an illusion?

But I didn't really believe that, nor did I think Paul had no memories. And what about Noel? My granddaughter had already revealed, through her drawings, that she had witnessed events so profound, yet so realistic, she was able to remember and document them. At Noel's age, it seemed unlikely she would have created this character and her relationship with "him" out of imagination or from something she'd read or seen on TV. It was certainly possible Noel could have seen a drawing of one of the smaller greys (though these beings were not so commonly depicted until recently).

Even if Noel's drawing proved to have come entirely from her imagination—or if it was merely a stylized representation of something else, something very earthly in origin—I could not just forget it. I could not continue to keep quiet about this high strangeness in my life. Something had to give!

But I couldn't face Paul and Sandra alone. I wasn't worried about how Sandra might react to this news; she had a curious nature and would probably be fascinated! But when I thought about Paul and his dispassionate approach to life, my resolve crumbled. Maybe my father had the right idea after all: Ignorance is bliss. Or is it? Could I be certain Paul was not already affected, even subliminally? I was going to have to do this, no matter how difficult or embarrassing it turned out to be. But I didn't have to face the music alone....

Anna was quick to support me in my decision, offering to go along if I wanted her. Of course I did! My family might *think* I was crazy, but they'd be less likely to say so in Anna's presence. I didn't much care what was said after we left; the admission would clear my conscience, the door left open for discussion.

We decided to plan our visit for evening, so that if things got uncomfortable we could excuse ourselves because of the late hour. I had forewarned Paul that I had something serious to discuss with both Sandra and him, but refused to divulge any more over the phone. Besides, I didn't have a clue what I would say once we arrived!

I fretted during most of the drive, arranging and rearranging a speech I knew I would never actually use. Would it be best to just blurt it out? *"Guess what? Anna and I have been abducted by aliens—I think you have, too, and your daughter as well! But don't worry, none of us are crazy; we only sound crazy."* Or was it better to draw Paul aside and speak to him privately? He was likely to be more candid in a one-on-one discussion. That seemed like the best route.

We arrived just after dark and parked in front of the apartment complex, but didn't immediately get out of the car. Anna put her hand on my arm, sensing that I was again having second thoughts about this confrontation. That was how I saw it, a showdown between a pragmatic young man and his confused but determined mother. Paul's philosophy maintained that if he couldn't see it or feel it or take a picture of it, it couldn't exist. My horizons had broadened, and with them my attitude: "There's more in heaven and earth, Horatio, than is dreamt of in your philosophy"—an overused quote, perhaps, but entirely appropriate.

The Christmas holidays were over, but a few diehards had not yet taken down their lights and decorations, which blinked and glittered through frosted windows as we walked by. For some reason these symbols gave me hope. I was doing the right thing for myself and for my family. It would turn out okay.

We were welcomed warmly, though Paul's expression defied his attempts at social niceties. He was curious yet ambivalent; wanting to know what this was all about while afraid we might actually tell him! There was no point dragging it out. I ushered him into Noel's bedroom, leaving Anna in Sandra's care. Noel was playing happily on the living room floor, ignoring us, but I was sure Anna would not divulge anything that might influence her. The decision to include Noel would be up to her parents alone. Neither Anna nor I would interfere.

I closed the bedroom door behind us and pulled up a chair, motioning for Paul to sit on the bed. I wasn't sure how to begin; all my previous attempts to prepare a convincing yet sane talk evaporated with his first question:

"You're not dying, are you?"

"No! Of course not!" I laughed, wondering why he would think such a thing. Simultaneously, I imagined Anna trying to convince Sandra that her mother-in-law was definitely not dying of some horrible disease; she was merely nuts. "Is that why you thought I wanted to talk to you privately?"

Looking embarrassed, Paul bristled, "You drove all the way down here, told us you couldn't talk about it over the phone, even brought Anna along for moral support—what was I supposed to think?"

He was right, of course. I should have suspected it. Did I think they'd automatically assume it had to do with alien abductions? I apologized for worrying them, realizing that this misunderstanding might actually make the real reason for my visit easier for him to accept.

It brought to mind something that had happened many years before. I had received my electric bill and was astounded by the amount due: $1,200! How could this be? Even during the worst months of winter my electric bill rarely exceeded $80. Irate, I phoned the electric company and berated the first person I talked to, who, of course, could not help me. I was passed on to the next poor unsuspecting employee who listened patiently to my tirade while furiously searching through computer records in hopes of finding an explanation. Finally the representative came back on, apologizing profusely for the company's error. It seemed they had inadvertently added an extra zero to the total, and that the actual amount due was only $120! I was so relieved! Even though the correct amount was much higher than normal, it was insignificant by comparison.

I dearly hoped Paul would find my story of alien intervention insignificant by comparison, too.

Taking a deep breath, I forged ahead, beginning with the first episode back in December of 1991 and continuing through the most recent event, only a week before. Paul sat quietly through my monologue, showing no emotion whatever, his face a complete blank. A long moment of agonizing silence followed as I clenched my hands in my lap and restrained myself from prompting him. At last he sighed; a signal, I hoped, that he was about to speak.

"You, uh, and Anna—you've both seen these things, together?"

"Not always. Sometimes, though." I wanted to ask him if any of my descriptions sounded familiar, but I was afraid of putting ideas in his mind that might not really be there.

He paused, then asked bravely, "And you said Granddad was involved, too?" I nodded. "I know it must have taken a lot of guts to tell me about this stuff, Mom, but you can't blame me for being a little incredulous."

I certainly didn't! But at least he hadn't laughed—not aloud anyway. Had he heard anything about this phenomenon before, I asked?

"Yeah. I think most people have. And if it will make you feel better, I really do believe UFOs exist. There were a lot of reports during the Gulf War of flyovers and some close calls with our fighters over there. Some of them even showed up on radar. These guys, these pilots, are sharp, Mom, and I don't think they'd misidentify something like that."

We were getting off track here. Paul sensed that I had something else to say and encouraged me to continue.

"You think since Granddad is involved, maybe we are, too, right?"

"I don't know," I admitted. "I do have some questions about when you were a boy, though, that maybe we should talk about."

"Like what?"

"Do you remember anything happening when you were young, anything that you still can't explain today?"

He thought that over a moment, then volunteered an incident that he recalled from our brief stay in Savannah, Georgia. Paul was four at the time and had been having problems sleeping through the night because of what he said was "cats staring at him through his bedroom window." (See Figure 12.)

"Do you remember what the cats looked like?" I asked, my heart pounding. I saw him in my mind the way he looked then: A towheaded boy, tall for his age, and strong-willed. I pictured him sitting bolt upright in his bed, screaming that the cats were going to get him as I watched in horror from his open bedroom door, helpless in the face of this familiar threat....

Paul at age 4.

Beth at doorway of bedroom.

Figure 12. A similar memory that both Beth and her son Paul have of his childhood terrors.

"Well, they were just cats, I guess.... Sort of cream-colored and shiny, probably short-haired. Usually three or four of them would be outside the window, lined up in a row, and just stare in. It was really unnerving for a kid to wake up and see something like that." (See Figure 13.)

"What did you think when you saw them?"

"That I was having a nightmare, I guess."

I sighed, wondering if I should prod his memory. What if my recollection of those events didn't coincide with his? Did that mean I had imagined them, or had I simply rearranged my memories to mesh with my own childhood terrors?

"...It's really vague, you know, after all these years."

Apparently Paul was looking for some confirmation from me, so I dared a leading question: "Do you remember what you did when this happened?"

"Oh, yeah! I screamed for you!"

"Did I come?"

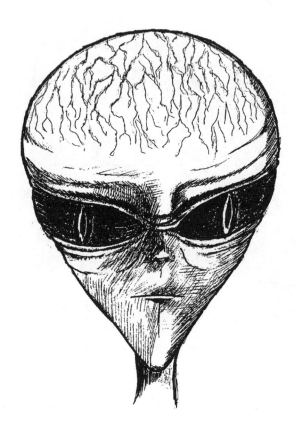

Figure 13. The "Cat Man," seen by three generations of Beth's family.

"Uh, uh," he stammered, looking suddenly uncomfortable, "I guess. Sometimes."

"What happened on those times I came in to see what was wrong?"

Paul looked stricken now. His complexion had paled and I noticed a thin line of perspiration had formed on his upper lip. Had I struck a nerve, exposed some deeply repressed memory he had struggled to keep buried?

"Nothing! Look, Mom," he snapped, "we both know that never really happened!"

"What do you mean, 'That never really happened?'" I asked, surprised. "You just said—"

"I know, but that never really happened!" he insisted. "It's just something I *think* happened because you asked about it!"

I hadn't, of course, asked about that particular incident. Paul had volunteered it. Even so, pursuing this topic was self-defeating and only served to upset us both. Changing tack, I told him I was more concerned for Noel and that she was the real reason I felt it important to tell him what had been going on. Paul scoffed at this notion, saying Noel was just an ordinary little girl with an active imagination.

"Then why are you upset by her drawings?" I inquired carefully.

"I'm not upset by her drawings!" he denied.

I disagreed, recalling how Paul had systematically removed Noel's drawings of *"Nu"* from her bedroom walls because, as he'd defended, "They gave her nightmares." But not wanting to upset my son further, I relinquished, only requesting that he and Sandra at least encourage her to talk about anything that might be bothering her.

On that note, we returned to the living room where Sandra and Anna were examining some of my own drawings we'd selected to bring along. Sandra had, as expected, found the phenomena fascinating, but frightening. Although Anna had taken pains not to mention my suspicions about Paul and Noel's possible involvement, Sandra had offered some interesting information of her own. She described waking up one morning to find several long scratches on Paul's back, scratches she was positive she had not inflicted. When asked about them, Paul had only shrugged, saying he didn't know how they'd gotten there.

"It scared me," Sandra admitted in a shaky voice. "It had happened before, and I just didn't know what to think! He couldn't have done that to himself—and I sure didn't do it!"

"I don't know anything about that," Paul mumbled.

Sandra stared at him, dumbfounded, but didn't challenge him.

Satisfied that I had said what I came to say—and survived it— we said our good-byes and left. Perhaps little had been accomplished by this visit, but I felt better anyway. My son may decide his mother was crazy, but that didn't concern me as much as it once had. They had a right to know what was going on, and I was exhausted from trying to keep it from them!

The ball was in their court now.

54. Anna

I decided it was time to adjust to the fact that I was being abducted. There seemed to be no other explanation.

I had been getting bits and pieces of memories back. I saw images of their eyes staring at me from about an inch away. I saw other humans on tables very clearly, but I didn't know who they were. I saw patches of gray wrinkled skin on an arm or a shoulder. I saw a ball of light floating around the room. I saw an eye peeking around my bedroom door, and then an alien head emerged.

I woke up with a scoop mark under my hair at the base of my neck, with a triangular hole above my knee, with a fingerprint bruise on my arm. I've woken up exhausted. My menstrual cycle is all messed up again. I've found puncture marks in my navel. I have memories of an anal probe and was sore for days afterwards.

I've had repetitive dreams of being in school, of being taught. I'm continually dreaming of constellations I can't recognize. I have dreamt of lots of people milling about while I carry a small ferret, with no tail, in my arms. I dream of moving, packing, traveling. I dream of the men in white coats again, coming to take me away. It feels good.

It's too much. I would have to accept what's going on before I dissociated into multiple personalities. It must be less exhausting to accept rather than to continue to deny. I will—for today.

55. Beth

It was odd standing over her like that. I wondered if she even knew I was there. A small grey was holding a long needle-like object in one hand, his other hand holding her hair to one side. I was told to touch her hand, to quiet her so that she wouldn't become upset. Then the long needle was inserted into the back of her neck just below the skull. Anna didn't react, though I believe I must have jumped. The needle was removed slowly while I concentrated on what might be in the syringe's chamber. If it was empty, something would have been injected; if full, something obviously removed. But the instrument was withdrawn from view so fast I didn't have the opportunity to look.

Another grey stood off to one side of Anna's bed, apparently in case there was trouble. I dearly wanted to be able to make some, but found I could only move my arms—very slowly—as if I were trying to dog-paddle under water. Within seconds of the needle's extraction, the two greys floated up to the ceiling in unison, like strange dancers in a Fellini movie. Details which had been blurred throughout the experience suddenly sharpened, and when I looked up the greys were gone.

Anna was sleeping soundly, her breathing heavy and slow. I wanted to wake her, but couldn't. I was exhausted, so tired it was as if I hadn't slept for weeks! I trudged back into my own room and collapsed on the bed, falling asleep immediately.

The following evening, after Anna arrived home, I asked her how she'd slept the night before. She was vague in her reply, indicating that she had been restless, but had no memory of anything having happened. I told her about my "dream" and asked if we could check the back of her neck. And there it was! A puncture in exactly the same spot! It was fresh, but not bleeding, and according to Anna, not sore to the touch.

Had it really happened as I remembered? If it really happened, and was not a dream, why didn't Anna remember? She had been in some sort of altered state during the event, either asleep or paralyzed or both, but I'd been in a similar condition when in their presence and still had some memories—even if only fragments. Was she blocking her memories instead? That was more likely the case, since she had frequently complained of severe and sudden headaches whenever she had tried to force recall. And yet I believed Anna was far less emotional over these incidents than me! When I did remember what happened, even in fragments, I tended to respond emotionally, feeling traumatized and overwhelmed. Perhaps this emotional response was the catalyst for remembering....

Many of these experiences *were* very dreamlike: Our shared "dream" of July 1992 (which was explored further under hypnosis in November) would never have been revealed had neither of us mentioned it! It certainly seemed like a dream, yet the similarities were striking, too many particulars the same to write the incident off as coincidental—or as some unheard of psychological phenomenon. The physical aftereffects were also difficult to ignore: muddy feet, dirty hands, leaf fragments and smudged dirt in our beds the next morning. A very realistic dream!

In the fall of 1992, I awoke about 2:30 A.M. for no known reason. I was awake for some time, listening. I expect I was waiting to hear Anna's brother, Rick, come upstairs to bed. The hall light was still on, which was odd since he never failed to turn it off when he came up for bed. He often stayed up late, not retiring much before midnight, but it was almost 3:00! Thinking he had fallen asleep before the TV, and knowing I couldn't go back to sleep while the light was on, I decided to get up and turn if off myself. Besides, I was wide awake and needed to go to the bathroom. I got up and opened my bedroom door. There, facing me at the other end of the hallway, was Anna's brother!

"What's wrong?" I asked him.

No reply. I just received a blank stare as if he were sleepwalking.

I repeated my question, worried that he could indeed be sleepwalking and I might startle him. He was very close to the top of the stairs, so if I did wake him suddenly, he could tumble down the steps before I could reach him. Fortunately, he stood motionless, seemingly unaware of me.

Abruptly I realized I was completely nude! Where was my nightshirt? I'd been standing under the hall light in my birthday suit—for God knew how long—trying to get this man's attention at three in the morning! Was this a dream? It was more like a nightmare! Before I had time to react, Rick suddenly turned and walked into his room without a backward glance or any indication he knew I was there. With equal aplomb, I went on into the bathroom, then looked in on Anna, hoping she had not witnessed my exhibitionism. But she wasn't in her room! The hall light shone across her room, clearly illuminating her bed, but she was definitely not in it. After double-checking by pulling back the covers, I went downstairs to see if Anna might have gone for a late-night snack. She wasn't there either. Instead of searching further—or worrying about her mysterious absence—I returned to my room, switched off the hall light and went back to bed, falling asleep quickly.

The next morning I awoke hearing my alarm, but couldn't seem to reach it. My head was at the wrong end of the bed! Struggling to right myself, I hit the snooze button and crawled under covers. I was freezing! When the alarm sounded again, I reluctantly crawled out of bed—and the memory of that early morning confrontation with Rick assaulted me. But I had my nightshirt on! It was inside-out, but I could have done that in my rush to put it on before going back to bed, couldn't I? Believing that the incident was only a bad dream, I decided to forget it.

I fed the horses on schedule, though I was unusually tired and so ravenous I could hardly wait for breakfast. Normally I didn't eat breakfast, preferring instead to just have coffee before feeding the horses and then an early lunch. But on this particular morning (as with so many others after an experience), I cooked a full breakfast. I decided to eat with Rick, subconsciously looking for confirmation that the events of the night before had never taken place. When nothing was mentioned, I began to relax—prematurely, as it turned out.

As we took turns washing up our breakfast dishes, he asked me if I would look at two holes he'd found in his hand. They hadn't been there the night before, he claimed; he'd found them when he was washing his hands and they began to bleed. I looked closely at the rather large puncture marks on the top of his hand, but hesitated to suggest where he might have gotten such an injury. He was watching me expectantly, perhaps assuming I would automatically attribute the marks to alien interference, but instead I asked him what time he'd gone to bed the night before.

"I don't know.... Late. I think I fell asleep in the chair," he guessed. "Why?"

I shrugged, content that what I remembered never really happened. Then, to my surprise and horror, he added suddenly that he'd had a weird dream of oriental men staring at him while he slept. Upon awakening, he saw a shadowy figure at the end of his bed, but it disappeared and he went back to sleep. Although he had of late been more forthcoming about his memories and descriptions of vivid dreams, he had resisted any attempts to explore these in detail. Instead, he would usually ascribe them to lucid dreams or beer-induced fantasies.

Anna, on the other hand, had also awakened very tired and hungry. She had suffered mild diarrhea that morning, but only remembered having dreams and tossing and turning a lot.

Another dreamlike incident involved Bob Huff's wife. The dream was memorable, she had reported, because I had been included. In this dream, we were walking single-file across a field or meadow, swishing through tall grass that came up to our waists. I was leading, apparently privy to our ultimate destination. She didn't seem to know where we were going or why we were walking through the field, but felt an unexplained urgency nonetheless.

Oddly, I had experienced a similar dream the night before. I remembered slogging through waist-high grass, yellow and dry to the touch like hay past due for harvesting. Wearing only my purple nightshirt and no shoes or slippers, I knew I was dreaming, but my senses seemed particularly acute. At first I thought I was alone, then heard the sound of grass being pushed aside behind me. I turned to see who it was and recognized Bob's wife immediately. Nodding, as if relieved to see her there, I continued on. Upon awakening, I was able to remember up to that point in the dream, but the rest was lost.

Curious about these similarities, I asked her if she recalled what I had been wearing in the dream.

"Some sort of shorty nightgown, I think." Was I wearing shoes? I asked. Was she? "No. We were barefoot."

"If I was ahead of you," I probed, "how did you know it was me?"

"Because you turned around once and looked at me," she replied matter-of-factly. "I thought right away it was you."

Another shared dream? Was it, in fact, an ordinary dream with a common theme, one many people experience but have no reason to discuss with others? Or could it be a dream memory instead, the mind's way of stimulating repressed memory through dreamlike images? I didn't know, but it bothered me. It concerned me, too, that my mind might opt to release repressed memories by way of trickery! How was I supposed to sort things out if I couldn't determine whether the memory was mine through dreams or everyone's through the collective unconscious!

Suppose Jung's theory of the collective unconscious was true? What if the whole phenomenon was in the human genetic code from the beginning of time? No. I didn't really believe that. It was too convenient—and it didn't explain the scars, which were certainly not stigmata, or the selective power outages, or our conscious memories of childhood events. Given those mysteries, how was I to attribute these "shared" dreams? Where did they fit in with the other unknowns?

Maybe they didn't fit in. Maybe they were a phenomenon all to themselves.

Dreams aside, Anna and I continued to be abducted from home and from other locations. If one of us was taken during the night, usually the other was taken, too, even if we did not see each other during the abduction. But always the memories came only in bits and pieces, as if full conscious knowledge of these intrusions was to be denied us forever.

56. Anna and Beth

When would it ever stop? It seemed that almost every day something strange happened! We suspected paranoia accounted for a large portion of our anxiety, but that didn't do much to calm the fears.

We had both learned to tolerate the lesser misgivings; driving alone or after dark, discovering small but previously unnoticed marks on our bodies, feeling watched. Even so, we frequently experienced panic attacks for no known reason. Often these attacks were quite severe and unexpected: One of us might have a sudden urge, for instance, to deviate from the standard route of travel onto a side road, a route that would take us far out of our way. These urges would almost always be preceded by feelings of being watched and then subconsciously directed to change course. When the pressure was denied, intense anxiety followed. These attacks seldom lasted for more than a few minutes, but were nonetheless very alarming and cause for concern.

Another cause for concern—and one just as frustrating—was a rise in the number of physical maladies. We had both experienced illnesses which had confounded our doctors: A diagnosis of leukemia for Beth (which was found to be nonexistent after she received a third opinion); a possibility of gall stones with Anna (which after numerous scans and tests proved not to be the problem); changes in eyesight (which have gone unexplained and uncorrected since 1992); and digestive/dietary disorders for both of us.

The frequency of abductions combined with bouts of depression and dietary upsets resulted in irregular sleep patterns. Afternoon naps did little to alleviate this sleep deprivation, and we soon developed problems with concentration and coordination, directly related to the interruption of REM sleep. It was during this bleak time that we experimented with self-hypnosis and auto-relaxation techniques in hopes of regaining the required sleep. This has helped, but as the intrusions and other problems related to them have continued, the process has not been entirely successful; respite is brief and often irregular. We keep trying.

The most upsetting medical anomaly, though, was the periodic pregnancy symptoms. We have both experienced these missing pregnancies at different times in our lives, and have been unable to attribute these symptoms to something explainable or commonplace.

Beth

I had two confirmed pregnancies that did not go to term (nor were they terminated by miscarriage or induced abortion). One episode occurred in my teens and has been described earlier in this book; the other soon after my marriage while my husband, who was in the military, was away on duty assignment. In this instance, I had gotten up in the morning with severe cramps and nausea. Feeling I was about to throw up, I rushed to the bathroom and then collapsed.

A period of time passed (estimated to be about two hours) when I came to and found myself huddled on the floor against the bathtub, with a trickle of blood on the tiles between my legs. Weak and unable to get up and go for help, I pulled a bath towel from the rack and used it to clean myself up. Some time later, a knock was heard at the back door. I called out, hoping I had left the door unlocked. My neighbor, who was the minister of my church where I sung in the choir, had come by looking for me when I hadn't appeared for choir practice, which had been scheduled for 10:00 that morning. Finding me in such a condition, he immediately went for his wife who called the military hospital and requested an ambulance.

After I was examined by two military doctors, I was told I'd had a miscarriage. I wasn't aware I'd been pregnant! I had been taking birth control pills since before I was married, and though I understood they weren't 100% effective, I had not experienced any symptoms of pregnancy. My periods had been regular, none skipped. The doctors couldn't believe I didn't know, arguing that I must have been about three or four months along at the time of the miscarriage. This wasn't possible, I knew, since my husband had been overseas during the time I would have conceived! I hadn't had an affair; I hadn't slept with anyone other than my husband!

They wanted to know where the aborted fetus was. Had I flushed it down the toilet? I didn't think I had ever made it to the toilet! I had no memory of what had transpired from the time I entered the bathroom until I came to squatting on the floor next to the bathtub, which was about ten feet from the toilet. (The house was over eighty years old, and as was common when such aged homes were refurbished and modernized, the bathroom had been added by appropriating portions of connecting rooms, making the bathroom oversized. This bathroom was approximately fifteen by twelve feet, large enough to accommodate a vanity and bench as well as a small dresser for storing linens.)

I couldn't imagine myself alert enough to crawl from the doorway to the tub, yet not conscious enough to pull myself onto the padded bench—which was certainly much closer! If I had crawled across the room, why hadn't I gone the other way, back out into the hallway where I could reach a phone and call for help? I tried explaining this to the doctor, but he would have none of it. I had obviously flushed the fetus, he insisted. It didn't matter, he assured me, how I had done it or that I didn't remember doing it. The fetus had not been found, so it was a moot point.

After several days in the military hospital with no treatment forthcoming, I called my father in Virginia. Although my husband had been notified and was on his way home, I couldn't wait for him to help me. My father arrived the following day, spent some time discussing my situation with the doctors, coming away visibly annoyed; they had not scheduled any treatment and we both felt a D & C was in order. I was in extreme pain and still bleeding. He made immediate arrangements to have me moved to a civilian hospital fifty miles to the north where I was thoroughly examined, given a D & C, and released the following day. The attending physician had warned me that complications may arise from the delay in treatment, but that there was nothing further he could do until such time as symptoms became apparent.

(Note: In 1972, at the age of twenty-six, I had a complete hysterectomy after years of irregular and painful periods. There was considerable scar tissue found in the fallopian tubes and cervix, and a number of small benign tumors on both ovaries. These, the doctors diagnosed, were the result of numerous complications during pregnancies, and also from several abortions and/or miscarriages. Having never had an abortion and only knowing of two possible miscarriages, this explanation was hard for me to accept at the time. Even so, I clearly recall being relieved that I could no longer become pregnant! I felt I had been pregnant too often in my young life, and looked forward to a future free from the possibility! I had a perfectly healthy son and didn't desire more children—especially children I was never permitted to see.)

Hysterectomy or no, I ushered in 1993 with definite signs of being very pregnant! After an abduction in late December, where I recalled a painful procedure having been performed through my navel, I awoke to find blood on my sheets. Examining myself carefully, I saw a large red-rimmed gash just above the lip of my navel. I knew I'd had a similar mark there before, but nothing unusual seemed to have come from that incident—just the hole, which eventually healed leaving a very small scar. Using a mirror, I looked more closely at the wound: It was in the exact place the other, smaller one had been, but this one was about a half-inch across and still open. When I pulled the wound apart I could see it was quite deep, though I couldn't see how deep because of the folds in the navel. The wound seeped for several hours afterward.

I wondered if I'd ever retrieve more memories of that abduction, but by the new year nothing more came and I soon filed it away in my journal as just another experience.

By the second week in January, I found I couldn't button the waist of my riding breeches with anything short of a wrench! I had worn that same pair only a few weeks before; they had been washed countless times and had never shrunk, and I knew I hadn't gained any weight. That morning I had felt queasy and had trouble keeping food down; perhaps, I thought, I was coming down with something and the bloating was only gas. But by late afternoon I was feeling like myself again—even if my pants were still too tight!

As the days passed, I realized the morning sickness was not going away, and my stomach had become so enlarged that my old jeans—which had long since been relegated to the dust bin because they were too large—were the only pants I could squeeze into. Compounding the problem, I was obsessed with house cleaning: washing floors, waxing furniture, vacuuming, dusting—even washing windows! I had never been a big fan of housework, suffering through it only because it was occasionally necessary in order to find lost items. And I couldn't stop! This behavior went on for more than a week, a phenomenon I recognized as "nesting." I was quite familiar with it, having experienced the same "nesting behavior" when I was pregnant with my son. But why was I acting that way now? I couldn't possibly be pregnant!

I eventually came to the conclusion it was all in my mind—or a symptom of some other physical problem. But just to be safe, I decided to purchase a home pregnancy test. In the morning I used the test, but it was inconclusive. Determined to get to the bottom of it, I told Anna I was going to see a doctor the next day.

That night I was abducted from my bedroom and when I was "floated" back, the little gray shits dropped me from three feet up! I fell, hitting the floor several feet from my bed and banging my elbow painfully against a wooden stool. This dramatic entrance seemed to have jarred my memory: I saw myself lying prone on a table, two techs standing on either side of my spread legs, leaning over me. Doc was to my right, his hand on my shoulder. Suddenly I felt a horrible stab of pain in my abdomen that subsided only after Doc placed his hand on my forehead, much as he had done during previous encounters. I was not told what had been done or why, despite repeated demands that I had a right to know.

Feeling helpless and depressed, I spent less and less time around the horses, planting myself in front of the TV in the middle of the day, a sedentary position to which I was not accustomed. One afternoon as I ate lunch in front of the TV, not really watching the talk show in progress, I heard the term "artificial insemination" and was instantly alert. The interviewer was pointing the microphone at a middle-aged woman who was recounting her recent experience with insemination. It seemed she had a necessary partial hysterectomy some years before, but not having had any children, decided to risk a revolutionary procedure: insemination.

Some problems were evident, but these were minor, the woman said, adding that the most frustrating part was not knowing if it had worked! When artificial insemination was used, home pregnancy tests would not show a positive result even if pregnancy was confirmed. Happily, though, the procedure resulted in the birth of a healthy baby girl. (But artificial insemination was not, to my knowledge, a new and untried procedure.) My sentiments were repeated by the host, who was told that it was revolutionary because the mother no longer had a uterus and had lost one ovary. The fetus, of course, could not be carried to term and delivered in the normal fashion; it had to be delivered by C-section.

An authority on the procedure was next interviewed: The surgeon explained that this technique had even been tried with a woman who had no reproductive organs whatever, who had undergone a radical, complete hysterectomy. Where did the mother carry the growing fetus? the surgeon was asked. The fetus attached to the abdominal wall, the audience was told. Although the procedure did not result in a live birth in this particular case, the doctor admitted, with more time and research it was certainly possible that any woman who wanted children could have children, even if all reproductive organs had been removed.

"Oh, God!" I said aloud. Did this mean I could have been pregnant? Having no reproductive organs no longer appeared to be an obstacle—and although I couldn't *get* pregnant, I might have actually *been* pregnant, through insemination. If *we* had the technology now, how long might *they* have had it?

I sat before the TV, no longer listening to what was being said on the show. I thought back to another abduction the year before when I was shown a pale, emaciated female who was holding the hand of a tiny, similarly weak-looking child. I was told by Doc that the little one was mine through *her*, its mother. I immediately denied them both, refusing to accept that these sickly creatures were in any way related to me. And I still felt that way!

It didn't make any sense for them to impregnate me, when there were so many women out there who were perfectly capable of bearing children! "Why go to all the trouble?" I asked Anna later that evening.

"Maybe because you wouldn't suspect anything—certainly not that you might be pregnant!" she offered. "You wouldn't be in any rush to see a doctor and have it confirmed either, would you?"

I had to admit she made sense—if any sense could be made from all this craziness. But surely it presented them with a few special problems, I argued. It seemed a lot simpler for them to bypass women like me in favor of millions who had all their parts! Besides, having it attached to the abdominal wall could not possibly imbue it with my genetic code! I didn't produce the egg, so the fetus would get nothing from my genes....

"But if several generations are being taken because of genetic experimentation," she went on, "and the aliens had the technology to create offspring for their own obscure purposes, why couldn't they also have solved that problem? We don't even know if it *is* a problem for them. I mean, it looks like if you're in the program, you're in it for life—and so are your progeny."

Or for as long as you were needed.

Anna

I seem to be preoccupied with maladies of the body. I don't think this is what the therapists meant when they told me it was healing to the mind to get in touch with how my body was feeling.

I was feeling pretty lousy with all the digestive upsets, irregularities with my period, and lately had developed a tremendous backache. I remember being told, "You must be taller." Sure enough, soon everything looked short. Tables seemed low. I could see over the partitions at work, my sweaters seemed too short. I expressed my fears to Beth: "You won't believe this, but I think I'm growing taller." So, she put me against the wall where we have, for years, measured everyone in the family. I was a half-inch taller than I had been in 1985! No wonder I had backaches.

With this new information, and the continual backaches, cramps, ovarian pain and spotting instead of having a real period, I made an appointment with the gynecologist. As I was being weighed, I asked the nurse to measure me. She measured me a full inch and a half taller than I had been! I was shook up when I saw the gynecologist and after the examination when he told me there was nothing wrong with me. I blurted out that I thought I'd been abducted by aliens and that they were causing my problems. (I was tired of lying to everyone and trying to come up with other reasons for feeling bad all the time.) Maybe he could help.

What a mistake! His first question was, "What day is it?", followed by, "Who is the President?, Who is the Vice-President?, Do you remember what we talked about last time you were here?" I was lucid. I knew the answers to his questions. He was asking me some of the questions from the Short Portable Mental Status Questionnaire, an indicator of dementia or reduced mental functioning. He didn't believe a word I said about alien interference, although he did take blood to check my thyroid (negative results), and demanded a drug screen, with the nurse present as I gave her a urine sample (also negative).

The gynecologist basically ordered me to see a psychiatrist. He made the appointment for me while I was in his office. I was so upset that I didn't feel I could refuse any of his demands. The psychiatric evaluation turned out negative as well; she said I wasn't crazy, but she couldn't help me because her treatment protocols didn't deal with the alien abduction phenomenon. I never went back to see either of them.

I chose two internists to help me get to the bottom of my physical symptoms. The first one didn't find anything wrong after lots of tests, and I didn't feel completely comfortable with her. I didn't make the mistake of telling the second internist about my involvement with the gray ones. I had terrible pains after eating pork spareribs one night and he confirmed my suspicions that it may be gall bladder problems. The sonogram found everything normal, and he opined that I may have had one stone that passed in the four days between acute pain and the test. I should have been relieved. I wasn't. I wanted something normal, curable. But the pains subsided for several months. Maybe that's all it was.

I went back to the internist when I started getting pains in my back every time I ate something with fat in it. Gall bladder again, I suspected. Two grams of fat in any one meal seemed to be my limit before the pain returned. I was real tired of bagels, fruit, vegetables without butter, jam and toast with apple butter. It was fortunate so many fat-free products were available, but too bad they tasted so terrible. I couldn't eat a meal in a restaurant without having diarrhea, cramps or gas.

The internist ran blood tests for liver dysfunction—negative. I had a CT scan of my entire abdomen. The results? Unremarkably normal. We tried antibiotics for giardia infestation (a parasite that causes diarrhea) and for ulcerative colitis (an auto-immune disease). No relief. I'd lost about fifteen pounds on my restricted diet, and have since decided I will eat anything I want and put up with the symptoms—most of the time.

I have memories of the gray ones putting a patch on my chest (between the collarbones), and I watched, fascinated, as it was absorbed into my body. I remember being told that I must not eat fat because my fat-to-muscle ratio is out of balance. For whom and for what? No response. I liked my body the way it was!

The morning sickness, sore breasts and bloated stomach have come and gone again. It had to be the week I went on vacation! Three weeks later, my period returned. No, I didn't go to a doctor and I didn't do a home pregnancy test. I knew I was pregnant, but I didn't care. I just wanted them to come and get it, the sooner the better. I know I should have gone to the doctor. Researchers think that this will prove something, but it won't. Skeptics will always come up with lots of explanations for the crazy woman who lied about sleeping with a man, getting pregnant, and then miscarrying.

I've been asked why I don't go to the doctor when I know I am pregnant, or when I know I have had something injected into my body, or been made to drink some foul tasting stuff aboard a craft. Think about it. Even if I do remember an abduction, what's the chance of getting in to see a doctor the next day? What do I tell the doctor? "Please do a blood test, look for something strange. No, I can't tell you why or what you are looking for, do them all." Would I ask the doctor to look for elevated hormone levels, extra vitamins and minerals, or less of something that's supposed to be there? Or how about going to see a doctor every time I suspect an implant or bone intrusion? "Excuse me, would you mind doing an MRI so that I can check this out?" Do that a few times a month and see what your medical bills are. Besides, I'm not convinced doctors are as omnipotent as I have been led to believe. Every time I go to one, it's like a guessing game. At least now I have an internist who acknowledges his limitations. He says we'll keep trying things until the pains go away, admitting that sometimes he really doesn't know what's wrong. Maybe we'll get lucky.

(Note: We were not lucky. The internist was frustrated by these symptoms and has since released me from his care, offering to recommend another doctor. I declined his offer.)

No, I haven't become indifferent to my medical problems, but I'm learning to live with them. I still understand that something normal can happen to me, something that our doctors can diagnose and cure, so I get new symptoms checked out. I was concerned when I had missed two more periods (I didn't feel pregnant); after blood tests, the doctor said I was entering menopause. Hurray! No more chance of pregnancy! What a relief, no more babies! That lasted about an hour, then I thought about Beth's pregnancies. Oh well, I could still hope they'd stop doing that, at least for a while.

57. Anna

My current therapist wasn't achieving the results we had hoped for, so she suggested I try acupuncture. I was game; I would have tried almost anything to get rid of these headaches. It had worked for my horses, why should I be reluctant to try it?

My first session with Norris Blanks was a real eye-opener. Not only is he an acupuncturist, but he is also a therapist who used hypnosis extensively in his practice. I hit the jackpot. But he had a belief system I was not ready to embrace: He is a New Ager. He believes in the healing power of "the Light" and the "I Am That I Am"; that past and future life experiences impinge on us in the present, and that there is an all powerful spirit to call upon to help us in this life. It was pretty spooky stuff for me, but after all I'd been through, nothing was that foreign. He was not familiar with any of the abduction literature, didn't know about the gray ones, and felt that the majority of attaching entities were of "the Light" and worked for one's highest good. Not quite my opinion, but I was willing to work through this if he was.

We spent several sessions talking and exploring philosophies, usually for two to three hours at a time. He did a type of hypnosis where I relaxed, closed my eyes, listened to his voice, and conjured up images as he directed me. We mainly worked on my headaches, self-esteem, my fear of fear, empowerment and getting my life back in order. Only once did I conjure up a gray one; I made him turn his back and leave me alone! Boy, did I feel powerful!

After a few sessions, Norris asked me if I thought that I had ever made a contract with these energies (that's what he calls the gray ones), permitting them to abduct me. Of course not! I told him I'd heard of other abductees who thought that they had, but I didn't feel that way. He asked if I was willing to look for a contract and if I did make one, would I be willing to break it. I immediately said, "Yes!"

During our next session we talked about the contract and explored the distinct possibility that I may not be willing to break it quite so flippantly. I feel that they may have given me support and possibly love when I needed it. I remember crying myself to sleep at night as a child, feeling my parents didn't love me. Even if the greys caused those feelings, I somehow felt dependent on them. I didn't like feeling that way, but it was there. I hated them for doing that to a defenseless child. So yes, I wanted to know if I made a contract so I could decide what to do about it with more awareness of the consequences.

During the session on July 27, 1993, we talked for about an hour, then did a relaxation session where we worked on fear, and fear of fear. Then we decided to do a hypnosis session. I knew this would be a very different session than the ones I had done with Budd, but I felt comfortable with Norris. Budd Hopkins, as an investigator, was careful not to lead me to any preconceived conclusions. Norris, as a therapist, had no such need. Some of Norris's questions were deliberately leading, and I have no doubt that I confabulated during these sessions. It was hard for me to sort out what I really experienced from what I may have produced to please Norris. In some ways it didn't really matter; this wasn't straight investigation, it was primarily therapy. He never did convert me to the New Age philosophy, yet he helped restore my self-esteem and gave me access to information that helped me cope.

Before hypnosis, we talked about where I imagined my time-line to be; past, present and future. He asked if I could separate myself from the time-line and step out of it. I could. After I was hypnotized, Norris asked me if I had an implant (not a contract!). I said that I had an implant, put in the back of my brain.

Transcript of Hypnosis Session with Norris Blanks

Tuesday morning, July 27, 1993

Norris:	It is at this point that you have indicated that at some time not of this earth frame there had been an agreement to have an implant inserted in this life. Is this correct?
Anna:	Yes.
Norris:	Yes. But in this lifetime, you do not believe that the implant is for your highest good?
Anna:	Right.
Norris:	Were you deceived in the past to have this implant inserted?
Anna:	No.
Norris:	The agreement was made of your free will?
Anna:	I think so.
Norris:	Can we journey back to the time when the agreement was made?
Anna:	Yes.

Norris: You know you are secure, no harm can come to you. But we will go back and investigate, and find out the beginnings and the origins of this situation. What I would ask you to do now is to move back from your time-line. Now I would like you to move along the time-line either to the future or to the past, whenever this event occurred. Do you know whether this has come from a future incarnation or life cycle or whether it has come from the past?

Anna: No.

Norris: What I would like you to indicate with your finger responses, is should we go backwards along the time-line to the past? Should we go forward along the time-line? Okay, we will go forward along the time-line. I want you now to move along the time-line until you are drawn to the point where the agreement for this implant was made. Tell me where you are.

Anna: In a big room.

Norris: Is it night or day? *[After a long pause:]* That's okay. Is it light in the room or dark?

Anna: *[Sigh]* Light.

Norris: Good. Who is in the room with you?

Anna: I see a large head.

Norris: And do you see yourself?

Anna: No.

Norris: But you feel your energy and your presence in this room?

Anna: Yes.

[Note: I felt like I was a ball of light in the corner of the room.]

Norris: And is this room a place of instruction or learning or what?

Anna: It's a library.

Norris: Is it a library that has information within it?

Anna: Yes.

Norris: Is this information in book form, or computer form or light form?

Anna: I think there's some books, but they're real old.

Norris: Is it familiar to you?

Anna: Yeah.

Norris: And there's another entity in the room with you?

Anna: Yeah.

Norris:	Would you please describe as best you can this entity? You mentioned that there was a large head. Is there any other detail you can add?
Anna:	It seems to have stripes on the side of it's head, and the front part of the head is very developed.
Norris:	What about the eyes?
Anna:	They're larger than mine.
Norris:	Do they have a color or do they have pupils and irises like we know them or are they all one color?
Anna:	I don't know.
Norris:	That's okay. Does this entity have a body?
Anna:	It has two legs that are thin, but they have muscles on them.
Norris:	What is the color of the entity? Does it have a skin color or a covering color?
Anna:	Pink.
Norris:	Are you able to look into the eyes of this entity quite comfortably?
Anna:	All I can see is the side of her head.
Norris:	Oh, I see. So you're not facing the entity. What feelings do you have of being in the presence of this entity?
Anna:	Respect.
Norris:	So you have respect for this entity. Is this occurring within our solar system?
Anna:	No.
Norris:	It is beyond that. Did you have any idea where this is occurring?
Anna:	The red planet.
Norris:	By the red planet do you mean possibly Mars?
Anna:	No.
Norris:	How many light years is it from the earth? Do you have any idea?
Anna:	23 or 27?
Norris:	Are you communicating with this being or you're just in its presence?
Anna:	I think it's supposed to do something to me, but I don't know what.
Norris:	Can you ask it what it's supposed to do?

Anna: I'm supposed to learn something.

Norris: Is it something that you will later use in your earth life?

Anna: Yes. But I'm not supposed to know what it is.

Norris: I understand.

Anna: I don't.

Norris: And so the knowledge will be activated in you at the appropriate time?

Anna: Yes.

Norris: In respect to your earth life in this cycle, what is your mission?

Anna: Change something.

Norris: Do you know what? Is it a structure or a consciousness change?

Anna: I don't know.

Norris: Are there other people that you are associated with here on the earth that are a part of this mission of yours?

Anna: Yes.

Norris: Now these are friends that you habituate with?

Anna: Yes, some I don't know yet, though.

Norris: At this time in 1993, you are not fully aware of your mission?

Anna: Right.

Norris: When do you believe that you will be made fully aware of your mission?

Anna: 1997.

Norris: So what preparation are you going through at the moment? Can you tell me?

Anna: Some sort of metamorphosis. I don't know what that means.

Norris: You said earlier you were unsure that what was happening to you in this life was for your highest good. Do you still believe this to be true?

Anna: Yep.

Norris: Why do you believe that what is happening is not for your highest good?

Anna: They've lied before.

Norris: I understand. So you believe that you may have been deceived. When did they first lie to you, do you remember?

Anna: No, but they hurt me; they told me it wouldn't hurt—a long, long time ago.

Norris: They hurt you physically?

Anna: Yes.

Norris: The people that, or the entities, that you have dealt with, are these representatives of this energy that you are in the room with?

Anna: No, they're somehow different.

Norris: Are they interfering with some other pattern that you have set for yourself?

Anna: I didn't set it.

Norris: If another pattern has been set for you, do you think that the greys are interfering negatively or they are part of this pattern?

Anna: I don't know.

Norris: Do you wish to break the agreement that you have made?

Anna: Yes.

Norris: Why do you wish to break the agreement?

Anna: Because I don't know what it is.

Norris: Is it wise for you to break the agreement?

Anna: I don't know.

Norris: When you have been involved with these other entities, have you ever felt a sense of love from them?

Anna: No. Duty.

Norris: Are they living beings themselves or are they automaton type beings?

Anna: No, they're living.

Norris: What do you think will happen if you break the contract?

Anna: I don't know, maybe they'll find somebody else.

Norris: What power do you believe you have in this situation?

Anna: Not much. I'm a pawn.

Norris: Are you aware that you may have great guidance systems around you, and other beings that you can call on, who can help you to understand?

Anna: No.

Norris: You're not aware of this, but you feel much more in control of this situation than you did a month ago?

Anna: Yes.

End Transcript

Now I didn't just have the gray ones to worry about, I had dealings with the "Pink Lady" from a red planet as well, an entity that implanted a "thought bomb" in my mind! (See Figure 14.)

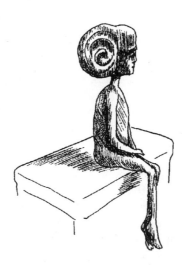

Figure 14. The "Pink Lady," who implanted a "thought bomb" in Anna's subconscious to be unleashed in 1997.

I didn't know if any of that was real, although she seemed real. I had accepted that I might have physical implants in my body at various times, as many other abductees had reported, but a mind implant? When Norris asked me what it looked like, all I could see was a small amorphous orange cloud. I now think that it was a memory implant rather than anything physical, that's why I couldn't describe it to him. I feel strongly that I do have knowledge in my head that will be revealed to me in 1997, but I sure would like to know what it is now. I can't quite buy the premise that it was implanted in my brain in the future. I may be willing to believe a lot of things that I never considered possible two years ago, but I hadn't lost all my analytical functioning.

I always feel better after the sessions with Norris. I'm more confident, more relaxed (and tired!), and feel more in control of my destiny. He's really helped me explore my feelings for myself, and my relationships with my family and the aliens. I guess that's the main goal of therapy: Be able to handle your own life independently of the therapist. I'd like to get there soon.

I was coping better, with Norris's help, but all we seemed to discover were more questions, not answers. I'd keep trying.

58. Beth

Fight back how? We tried alarms, sensors, cameras, witnesses—nothing worked! We'd had access to some of the best technology available, yet the shits kept coming! I felt as if we were trying to win a race by running backwards across the start line. I wasn't about to give up, no. That wouldn't accomplish anything. But what had all this technology done for us? There had to be another way!

Anna had read about other abductees who claimed limited success resisting the greys. One method involved chanting selected passages from the Bible; another promoted the use of mental resistance (refuse to go in your mind? Couldn't they continue to take your body?); yet another suggested tying oneself to an immovable object. There was one I found pathetically sad: Kill yourself. If you're dead, they won't want you anymore.

Had these people found relief, though? Well…not always. The abductions continued, but according to some, on a less frequent basis. Less frequent than when?

I have a friend, another horseperson and fellow abductee, who is a devout Christian. She believes in God, goes to church regularly, and tries to follow the teachings of her religious faith. During her many years dealing with these abductions, she has steadfastly clung to her beliefs, trying without success to resist them through faith in God. As with Anna and me—and many more like us—a belief in God, Jesus, the Virgin Mary or the Holy Spirit has not stopped the assaults. Chanting passages from the Bible or praying for relief do not seem to have any affect on these creatures. They have an agenda, a job to do, and although a higher authority may be privy to their plans for us, we apparently are not considered worthy of that knowledge.

It may be (although I doubt it) that these creatures report to the same higher authority as believers here on earth do, and if that is so, what does that tell us about our importance in the scheme of things?

In short, if there is a God and He (or She) knows of our predicament (as believers feel is true), then He (or She) has obviously chosen to stay out of it.

I think we are on our own with this one.

Conscious physical resistance has not worked either. One may *think* about clinging to an immovable object, but when *they* come to get you the body is effectively paralyzed and the grip is lost. This method of control works equally well should the victim show any aggressiveness towards them. I know because I've tried. When I was abducted in September of 1992 while on vacation, I was allowed to move in order to finish dressing myself, but as soon as I took a step toward them I was quickly immobilized again. Although this gave me a fleeting sense of power, it was nothing more than wishful thinking.

I had spoken to another abductee (while we were in New York visiting Budd Hopkins) who had actually struck one of the greys. She said the little fellow felt like he was made of papier-mache. But retaliation was swift—and strong. She was quickly knocked down. That, she assured us, was not likely to happen again.

Some diversions had been successful, though. These feeble acts of defiance did not usually stop the abduction, but they could sometimes slow the little buggers down. After one such experience, where they seemed dumbfounded by my polished toenails (asking me how I had injured my feet!), I devised a plan that might confuse them even more. Before going to bed one night, I wrapped each toe in a bandaid—not just your average flesh-colored bandaid, either, but those multi-colored glow-in-the-dark variety that are so popular with children. When I turned out the light, I checked my toes to see if the bandaids really did glow in the dark. Sure enough! It looked like my toes were radioactive! I was so pleased with myself it took me several hours to get to sleep. It was the only time I ever wished for them to make an appearance!

And they did come. And I did go. But when I woke up that morning, all the bandaids were gone! I searched my room from top to bottom, but never did find them. So maybe my defiant gesture didn't change the outcome, but I'd be willing to bet real money that it gave them something to think about.

Anna even suggested that I try dripping red nail polish on my skin to see if that brought a reaction. It couldn't hurt, so I tried it. But nothing changed. I expect by then they must have figured out what I was up to.

Although these little tricks did more to improve my outlook than to change my situation, it wasn't a complete loss. I learned I had not lost my sense of humor after all. I can still imagine those fluorescent bandaids hanging in an exhibition on board some craft, while groups of little gray shits cue up to see this marvelous and mysterious example of human cultural depravity!

And that was the key.... Humor could be a powerful weapon against fear and oppression. I did not like being a victim. I was tired of it! Maybe I couldn't be in control, but that did not mean I had to meekly succumb, either! Abduction against one's will is surely not a desirable state of affairs, but if I could just find humor in some of it— even at my own expense—maybe life on this plane would be easier. I might actually begin to enjoy things again. That was good advice, and I decided to make a genuine effort to follow it.

Closeup of
insectoid's hand.

Figure 15. An insectoid alien seen by Beth during an abduction experience. Neither Beth nor Anna have had personal contact with this being.

chapter 11

Assimilation

59. Beth

The year 1993 looked like it was going to be a repeat of '92—abduction followed by fragmented memory, followed by abduction, etc., etc. I found myself using Dr. Ruxer's coping strategies more and more. My day was filled with repeated questions about how I was feeling and why. I was working especially hard to keep a sense of humor, knowing it had helped so much in the past. But even that was becoming a chore.

I buried myself in my work hoping to alleviate some of the tension under which I lived. Anna and I had been at each other's throats of late, and keeping myself busy outside was the only way I knew to avoid confrontations. We understood why we were so antagonistic, but since our respective attitudes and belief systems were as different from each other as we were, it was difficult to establish a neutral ground where these disparities could be settled amicably. I was searching for alternate explanations while Anna was looking for confirmation. There seemed to be no middle ground here. Neither of us wanted to give an inch on our beliefs, even though we claimed those beliefs were always up for debate!

I escaped through work—it was easier than facing our communication problems. I believed that if I worked long and hard enough, my thought processes would eventually succumb to the body's demand for rest. But I worked almost as hard getting to sleep at night as I did being physically active ten to twelve hours a day! Often, after a particularly long day (one which might exceed twelve hours), I would collapse in bed then lie there fully awake until after daybreak. Even when the night was quiet—no visitations from the grey shits— it took superhuman effort to get up with my alarm and start the whole process over again, knowing that night would only be a repeat of the one before.

I knew I was tired, knew I couldn't keep up this torturous pace forever, but once started I couldn't seem to turn it off. It had become a habit; I was running on adrenaline most of the day, and when that ran out, I reminded myself that quitting early and returning to the house would mean having to face Anna. Would we argue over some trivial matter again? On a subconscious level, I understood my behavior, knew I was avoiding the issue, but I was on a roll and much too preoccupied with make-work projects to quit now.

In the past, whenever I neglected to take time to recharge my batteries, I inevitably came down with pneumonia. This had happened several times since my early twenties and there was no reason to expect that I'd not get it again. Yet I went on with this self-delusion until I could no longer perform even the simplest of tasks without stopping to rest. The sleep I forfeited, the long hours of physical activity, the concern over our deteriorating relationship—all these and more finally brought on pneumonia.

Adding insult to injury, I was forced to take time off from my daily routine to recuperate. I hated the inactivity, but more than that, I dreaded having to face Anna—with no escape. She had graciously taken time off work to relieve me from my duties, but needing only to fill in for the required morning feedings (since our barn help would be there during the day), this left plenty of time when we would be together. Avoiding each other was impossible. I couldn't hide in my room all day! Sooner or later we were going to have to face each other and work things out.

So we did. Not all our disagreements were settled—not by any means—but we were making a stab at it. We had to put things in perspective, recognize what was important and what was not, what required honesty and straight-talk, what was better off left unsaid.

It was a difficult process requiring many long hours of candid discussion. As hard as it was for both of us, we did start to feel more comfortable with each other again. After all, Anna and I had known one another a long time and this familiarity—as with siblings who, after years of separation, are once again living under the same roof— had threatened to breed contempt. We knew ourselves well; we knew each other even better. If we were going to get along, we would have to set aside our individual gripes and petty annoyances. We were dealing with something far more serious, something that should take precedence over these insignificant problems.

We should be thankful! So many people out there were trying to handle similar experiences with abductions and had no one in whom they could confide.

It was past time we realized how fortunate we were to have each other.

60. Anna

I'd decided I had to get a life. I could no longer continue to feel sorry for myself, and allow feelings of victimization to overwhelm me. I needed to start functioning as a normal person again. I couldn't do that without having a better understanding of my past. Once again, Norris Blanks agreed to help me do this. For me it was necessary; for him it was an adventure into the unknown. He'd begun to read several of the books I brought him from my library, and I began to read some of the books he suggested. I had objected to training another therapist, but this was different—we were both learning and exchanging philosophies. He was willing to put in the time with me, and I needed the help.

We scheduled another session so that I could explore more of my childhood memories. I felt that only by examining my past could I understand how to deal with the present. Norris felt that I had been severely traumatized as a child, and needed to understand that child's feelings, to release that trauma to become whole. What follows are partial transcripts of an hypnotic awareness session done on August 30, 1993.

Norris: We're going to go back and reassure that younger Anna, comfort and nurture that younger Anna and let her know that everything is okay. What I would like you to do now, is to travel back and drop in at a point where you feel the younger Anna really needed some support. How old is Anna at this point?

Anna: Ten.

Norris: What is Anna feeling at ten years old?

Anna: Scared. *[Crying]*

Norris: What is she scared of?

Anna: I don't know.

Norris: What is she telling you?

Anna: *[No response, crying]*

Norris: Nothing? That's okay. I want you now, perhaps to go back to an earlier part in her life where whatever she's scared of first originated. How old is she now?

Anna: Three.

Norris: Is she still scared?

Anna: No.

Norris:	What is she feeling at three years of age?
Anna:	Playing.
Norris:	Is it day or night?
Anna:	It's day.
Norris:	And where is she playing?
Anna:	White light.
Norris:	She's playing in the white light. Is it all around her?
Anna:	Yes.
Norris:	Good. Does she look happy?
Anna:	Yes.
Norris:	Is the white light her blessing and her protection?
Anna:	No.
Norris:	It is not. What does the white light represent?
Anna:	It's just sort of there.
Norris:	Can you describe the surroundings?
Anna:	She's all alone in the round room. Sitting on the floor.
Norris:	What's the floor made of?
Anna:	White stuff.
Norris:	We'll look around and consider this as part of an adventure, for you are an explorer. Do you understand?
Anna:	Yes.
Norris:	And there is no one in the room with her?
Anna:	There's one of those gray guys.
Norris:	Is there any communication going on?
Anna:	*[Sigh]*
Norris:	Nothing. Are you able to take her by the hand?
Anna:	Yes.
Norris:	I want you to take her by the hand and I want you to fill her with the great love that you have within you. And while you are doing this, I want you to convey understanding to her.
Anna:	*[Sigh]*
Norris:	What is she feeling?
Anna:	Scared.
Norris:	Okay. I want you to pass more love into her. For your love has the power to dissolve the fear. Okay?
Anna:	Yeah.

Norris:	Are you still holding her hands?
Anna:	Yes.
Norris:	Is she feeling any anger as well as the fear?
Anna:	No.
Norris:	It is just the fear. And she feels alone?
Anna:	No, now there's another little girl there.
Norris:	And what are they doing? Are they talking together?
Anna:	The other girl's holding her.

[Note: It was Beth.]

Norris:	Is the other girl older?
Anna:	Yes.
Norris:	What are you feeling towards her at the moment?
Anna:	She's protected now.
Norris:	Your energy protected her?
Anna:	No, the other little girl did it.
Norris:	How is she protected?
Anna:	She's holding her, and she's not going to let the other thing come near. *[Crying]*
Norris:	I want you to reassure her and I want you to fill her with your power and your strength and understanding. You're feeling very pleased with yourself because you have been able to look at something that you have not wanted to look at before, and by the very looking at it you are breaking the power that the fear held over you. I want you to look at the smaller Anna now. Tell me what she is doing.
Anna:	Sitting on the ground.
Norris:	How is she feeling now?
Anna:	Better.
Norris:	Are you able to go and hold her in your arms?
Anna:	No.
Norris:	Why can't you hold her in your arms? Is there a reason for that?
Anna:	The thing won't let me.
Norris:	Look at the entity and make the statement "I Am That I Am." Not in confrontation, but just as a statement of you and who you are.

Anna: She says I'm too little for that.

Norris: I beg your pardon?

Anna: I'm too little for that.

Norris: But, you're not too little. Tell him you're not too little for that.

Anna: *[Sigh]*

Norris: Say it again, I Am That I Am.

Anna: *[Sigh]*

Norris: What's happening now?

Anna: Nothing. It doesn't matter.

Norris: It doesn't matter. What doesn't matter?

Anna: It just doesn't matter. They've got me. I'm theirs.

Norris: Ask them why they've got you.

Anna: I'm supposed to be there.

Norris: I'd like you to tell him, that under universal law, you would like to be treated with the highest honor and respect.

Anna: [Sigh]

Norris: What is he saying?

Anna: They say they do—since we're property.

Norris: I beg your pardon?

Anna: He says I'm property.

Norris: You're property.

Anna: He does treat me with respect.

Norris: Ask him what is he doing? What does he want with you?

Anna: Wants to teach me things.

Norris: What sort of things?

Anna: Things for later.

Norris: What's happening now?

Anna: He gave her the block to play with.

[Note: This is the block that I turn inside out with my mind.]

Norris: Is she calm now?

Anna: Except when she throws the block.

Norris: Have all the traces of the fear gone?

Anna: Yeah.

Norris: Good. Is the other entity still there?

Anna: Yes.

Norris: Would you tell him that you believe that you have the right of choice?

Anna: He doesn't care. It doesn't make any difference.

Norris: Ask him when these arrangements were made.

Anna: A long, long time ago.

Norris: Before the beginning of humankind?

Anna: No, but a long time ago.

Norris: At what level is this particular entity? Is he at a high level or at the worker level?

Anna: It's someone who takes care of the kids.

Norris: I see.

Anna: Teaches the kids. Tests them.

Norris: What is he testing them for?

Anna: Powers.

Norris: What powers?

Anna: Mind. Mind powers, make the mind work in different ways.

Norris: And what will this achieve?

Anna: More able to work with them.

Norris: Will you tell him that you have the right to choose, and that you will choose?

Anna: He says that's okay. You can do that, but it won't make any difference.

Norris: Well, that is your decision.

Anna: He says it's okay.

Norris: Would you ask him if he understands love?

Anna: He doesn't understand.

Norris: He doesn't understand love. So he is purely an intellectual being?

Anna: Maybe.

Norris: Well tell him that you have the gift of love.

Anna: Yeah, he says that's why we're here. It's confusing.

Norris: That's okay. You don't have to understand it all. All you need to know is that you have activated some form of your own empowerment, and you are not a helpless victim of these entities. What's she doing now?

Anna: Playing with the block.

Norris:	What I'd like to do is for you to take her home, and tuck her in bed. Are you able to do that?
Anna:	*[Laughs]*
Norris:	What?
Anna:	This time she gets to stay in the top bunk. She doesn't get dropped on the floor!
Norris:	Will you tell me when you have taken her back home, tucked her in….
Anna:	Yeah.
Norris:	Comforted her, and made her secure. And I want you to reassure her in your loving way that there is no need for her to feel sadness or rejection, or loneliness, or any of those things. And that no longer will she be plagued by fear. Can you do that?
Anna:	Yes.
Norris:	Good. Where is she now?
Anna:	In bed.
Norris:	Is she comfortable?
Anna:	Yes.
Norris:	Good. Is there anything else that needs to be done for her, at this point?
Anna:	No, just let her sleep.
Norris:	So we will let her sleep and we will let her enjoy that freedom from fear. What I'd like you to do now is to come back to that period that you entered before when she was ten years old. I want you to come back and see if she has changed.
Anna:	She's cold.
Norris:	Why is she cold?
Anna:	Outside.
Norris:	She's outside.
Anna:	*[Shivers]* Shit.
Norris:	You have the power to look at it.
Anna:	They just brought her back; they left her outside.
Norris:	So where is she?
Anna:	Outside the door to the house. It's cold outside and she's in her nightgown. They shouldn't do that to a little kid!
Norris:	No, no they shouldn't. What's she feeling?

Anna:	*Mad!*
Norris:	She's feeling angry, she's not feeling poorly. She's feeling angry, right?
Anna:	Yeah. She's mad.
Norris:	Then why don't you let her be mad. Tell me what she's thinking.
Anna:	*[Crying]* Why do they always do this, and she can't tell anybody?
Norris:	I want you to get it out. I want you to be with her and let her get that anger out. That's it. Good. Let all that anger come out.
Anna:	*[Big sigh]*
Norris:	Let all that anger out. You don't have any need for it. Give her permission to get it out. Tell me when it's all out. Tell me when you think most of it is dissipated.
Anna:	She's tired.
Norris:	Good, but does she feel better now that she has released that anger?
Anna:	Yes.
Norris:	Yes, because she's been carrying it around for a while, hasn't she?
Anna:	They keep doing that to her.
Norris:	What? Dropping her outside?
Anna:	Yes.
Norris:	So, what have they been doing with her? She's ten years old, what's been going on?
Anna:	Hurt her belly. *[Rubs stomach]*
Norris:	Just tell her as she thinks about the pain, it will disappear. As she thinks about it, it will go away.
Anna:	Oh-h-h!
Norris:	Where is she feeling most of the pain?
Anna:	Around the middle. They told her she wasn't ready yet.

End Transcript

I was standing outside of the house in England. It was the house I lived in on the military base. I started my first menstruation while living in that house, just before I turned eleven years old. I believe the aliens had been checking my reproductive organs and that was why they said, "She isn't ready yet."

It was interesting to me that Beth and the inside-out block reappeared without any prompting. It was also nice to confirm my suspicion that I didn't fall out of a top bunk bed by myself; I had help. No wonder I didn't get hurt.

The question of humans being the alien's property had come up in my reading. The first person to mention it that I ran across was Charles Fort. It felt like a real statement from the aliens, not something I made up. Either way, I didn't like the implications.

I found these sessions with Norris very therapeutic, and I regained some memories as well. But his techniques were so very different than what I was used to with Budd Hopkins. It was hard to tell, in some instances, if I was recounting what the aliens told me or if my subconscious was responding with my own perceptions of things.

61. Beth

The day started as any other: I got up, went to the bathroom, prepared to brush my teeth. I squeezed the toothpaste onto the brush, jammed it into my mouth as I slumped over the sink—then I was somewhere else! There were two men standing before me, both tall and well-built, both blond. I didn't recognize them, didn't know where we were or why these men were staring at me!

"Listen carefully and remember this," one of the men announced. *"You will be going north, going where you must go...."*

Suddenly I was riding in the passenger seat of Anna's car, looking out over familiar terrain. I glanced over and saw Anna behind the wheel, apparently unconcerned and looking almost bored. We had taken this route many times on our way to the annual horse convention in Ohio, and I wondered if that was where we were headed. But something was wrong with this picture: The car was not cluttered with suitcases, boxes, discarded coats or the rest of the paraphernalia which usually accompanied us on our trek north each November. The trees had buds (some already in bloom) and we were dressed accordingly in sweat shirts and jeans; the windows were cracked to permit the flow of fresh air. The air smelled sweet and slightly damp.

We drove on. Our eyes met periodically, but nothing was said. We didn't need to talk; we understood what each was thinking. This silent communication did not feel unnatural or unfamiliar, and although I was worried, not understanding what was going on here, there didn't seem to be any way of stopping it. I thought about lighting a cigarette, then realized I didn't want one. Besides, we didn't bring any along—which was unusual in itself! Had we intended to drive so far without the comfort of our mutual habit to pass the time?

As an alternative, I switched on the radio, but heard only static. Then a man's voice, scratchy and sporadic, interrupted the radio's white noise:

"They're on the move! Keep...not too close.... Going to meet...sisters.... Stay...on—"

What did this mean? Had I tuned into a station that was out of range? Somehow I knew that wasn't it, that wasn't what we were hearing at all. Anna and I looked at each other, then stared ahead, accepting what we heard as part of the plan, whatever that plan was.

Then another voice was heard, this one in my head, as if I had become the receiver in place of the radio.

"Do not stop for any reason. If you are intercepted, drive through! Do not stop for anything or anyone! Do you understand?" (Anna and I nodded in unison.) *"You will meet your sisters in the north. You must arrive at the right time, when the time is right. You must not be interfered with; you cannot be stopped from reaching your destination! Do you understand?"* the voice asked again.

And as before, we nodded.

Foamy drool was running down my chin, the toothbrush dangling from my fingers. I shook my head and attempted to straighten up, feeling achy and stiff as if I'd been bent over for hours! What had just happened? Wiping the mess from my mouth and chin, I staggered backward and sat down on the toilet, shaking violently. I reluctantly let the images run through my mind, like watching a rerun of a movie I hadn't wanted to see the first time.

Was it a memory flashback? An induced fantasy? Had I flipped out at last? What did it mean? It had felt real. I could still smell the humid air blowing through the car's windows as we drove along...the static on the radio interspersed with the scratchy sounds of a man's voice...then the other voice in my head, the one that demanded we not stop for anyone or anything....

Was it meant as some kind of warning, an omen of something that might happen in the future? I sure hoped not!

I had been shown images during a previous abduction of women Doc had called "my sisters." And here it was again, another reference to the mysterious "sisters." It had to be in my mind, something I made up to clarify the original memory, to give it a reason for being.

Not knowing the significance of these "sisters" might have so frustrated me that I had compensated for it by making up this little drama.

That had to be it, I decided, shrugging off the incident as the mind's way of coping with this strangeness. It would probably never happen again. As long as I identified it for what it really was, it could be put away and forgotten.

Life went on its merry way and I had pretty much hardened to the eerie encounters and flashbacks, believing that the aliens lied regularly for their own obscure purposes. I refused to be taken in by their lies! I refused to follow their instructions, no matter how forcefully they were given! Besides, *contactees* received messages, followed orders, believed the aliens were Space Brothers only trying to help humankind—I didn't. I had seen no evidence that they wanted to help anyone but themselves!

I thought about John, the huge human-looking man who had appeared during several abductions, and wondered what purpose he and his like served. (See Figures 16 and 17.) John was brought out whenever I had supposedly ignored their instructions or orders, seemingly as an enforcer of some kind. He was intimidating. At about seven and a half feet tall with unbelievably broad shoulders and an imposing militaristic bearing, I never failed to cringe at the sight of him. Although he never touched me, his bland expression and severe manner bespoke authority and I was afraid of him. There was no room for argument or empathy in this man's world. Like the greys, he had a job to do and it appeared that job included intimidation.

In one instance, John (who had identified himself as leader of the security force—whatever that was) and two of his henchmen appeared after I had refused to hold an infant that did not look completely human. Again, neither he nor either of his sidekicks took any action to enforce the order, but I had worried that they might! John did speak aloud; his mouth moved and I could hear and understand clearly what he said. I simply refused to *do* what he said!

I had survived the onslaught of these "enforcers," taken them with a grain of salt, then ignored them—until an entirely new messenger came on the scene. I awoke one morning about 2:30 A.M. to a bluish-white light pouring through my bedroom window. I attributed it to moonlight before realizing that there was no moon that night. Then a shadow materialized in front of the window, as if it had been *poured* there, becoming more solid by the second. The figure was silhouetted by the light from the window, its features indiscernible. Looking to be well over six feet with extremely broad shoulders and narrow hips, yet with a very small head, I thought it looked too much out of proportion to be a real person. It had to be a dream—or perhaps a shadow cast by the curtains.

John, Unit Commander, Security Group (about 7 feet tall).

Pouch of some kind? →

Insignia? (over left collar bone)

Figure 16. John, the "enforcer."

Just as I had accepted the latter as fact, I heard a male voice: *"Talk to your sisters, the other mothers who are your sisters. You must talk to all the sisters and tell them to prepare."*

How bizarre, I thought! What a crazy dream this was! I went directly back to sleep.

Figure 17. Details of John's face.

But this character would reappear more than once over the next year, each time bearing a message that didn't seem to register as either timely or rational! One such message was: *"Do what you have been doing. Do not change your mind; do not waver. This is the right thing"*; and another: *"Tell all the mothers of two year-old children to go to North Carolina—to the coast."* These were repeated several times. The omen-like messages were always delivered in a nonthreatening manner, almost like a recording. The figure's image, which formed slowly, delivered its message, then faded out, resembled a projection of some sort—a hologram. No doubt it was this impression, so dreamlike, that fascinated me rather than frightening me—with one exception, so far.

The stables had always been my retreat, my safe haven from all the madness surrounding us. The gray shits had never intruded upon this sanctuary, and so I could relax there. One evening, while warming myself in the tack room waiting for the mares to finish their evening meal, I heard a commotion in the aisle outside. Assuming one of the barn cats had startled the mares into flight, I hurried to the door and peeked out. (The horses are fed in tie-stalls, narrow enclosures open at one end and just large enough for one horse to turn around. There are twenty of these connecting stalls running along one side of the barn, each fitted with feed bins and hay mangers.)

When I looked outside, I saw that most of the mares had escaped their stalls and were running up and down the aisle in near panic. Concerned for their safety, I hurried out, not bothering to put on my boots or coat. I grabbed a whip as I left, hoping to round them up and steer them back to their stalls—or chase them out of the barn, whichever worked best. A heavy chain was strung across the aisle to keep the mares from wandering down to the stallions that were stalled closer to the front entrance. Worried that the panicked mares might entangle themselves in this chain, I rushed to cut them off. When I reached the chain, the mares suddenly halted, their ears pricked forward, bodies tensed. I turned toward the front of the barn, trying to identify what had gotten their attention so suddenly—and I saw him. I sank to the damp ground, completely limp.

He was tall, over six feet, wearing a silver or gray uniform with heavy shoulder padding. The uniform, or suit, was one-piece from neck to toes and had no ornamentation of any kind. The man was handsome: Blond hair and blue-eyed with a fair complexion. He looked like something out of a fairy tale! Subliminally, I was aware the horses were still there, but frozen in mid-stride—as I was—but at that moment I didn't care.

In a deep voice—a voice I immediately recognized as one I had heard a number of times from the shadowy figure in my bedroom—he said, "Tell all the mothers with two year-old children to go to North Carolina—to the coast." Then, after a brief pause, he added, "The sisters will come to you when it is time."

And as before, once the message was delivered the figure gradually faded out and I was again reminded of a hologram. When he had faded out entirely, the horses snapped to life, turning around voluntarily and leaving the barn in an orderly fashion. I got up from the dirt floor, the dampness having penetrated my clothes so completely that when I tried to brush my seat off, my hands came away muddy! How long had I sat there? I didn't know, but I was cold and wet and angry! How dare they intrude on my safe place!

I managed to finish feeding the rest of the horses, returning to the house after dark. Anna and her sister came home soon after. While telling Anna about my barn encounter, and the now familiar message, I broke down, more upset by the experience than I'd thought. The message was admittedly laughable, but the fact that they were now moving in on sacred territory was more of a threat than their invasion of our bedrooms!

Was this the same one who had previously delivered the bedroom messages? Anna asked me after I'd calmed down. I thought so. I had never actually *seen* the one in my room, but I was sure the voice was the same.

During our conversation, as we attempted to figure out this new interloper, I found I was becoming very confused and distracted. Perhaps I was just tired and overstressed. I had good reason to be! In checking the time, I saw that it was much later than it should have been, too. By my calculations, it should have been close to eight, yet the clock read eight forty-five! How could that be? Did I lose an hour in the barn? It couldn't have taken that long for him to deliver the message! I was certain I had not been abducted, not taken anywhere. I had not seen any greys or any sign they had ever been there. Other than being a little damp and chilled (from sitting on the dirt floor of the barn), I had no unexplained marks or odd memories and could account—I thought—for every minute of the evening.

The following day I was still in a state of confusion, hardly able to complete anything I started because I couldn't seem to remember what it was I was going to do! The horses showed signs of confusion also, refusing to go down to the front third of the barn, a few feet before the aisle chain. It was several days before they settled down and returned to normal.

How long might it be before I did? What was normal, anyway? What did these messages mean and why were they being delivered? Surely they didn't expect me to actually *tell* mothers with two-year-old children to go to the coast of North Carolina?!

So they were lying. I had already accepted that.... But wouldn't they know by now that I wasn't taking these messages seriously? Why did they persist in this charade?

I wished I had some answers—any answers! Even if the answers came from the greys, even if they were all lies, it would be something. I yearned to write more than one sentence in a row in my journal that did not end in a question mark!

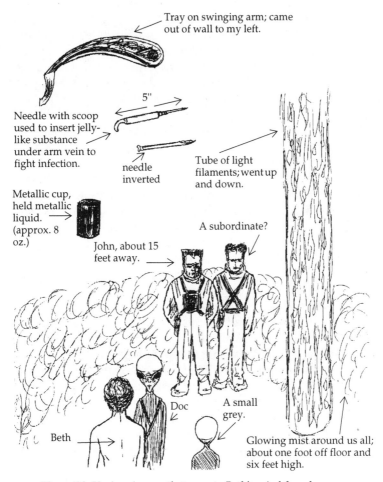

Tray on swinging arm; came
out of wall to my left.

Needle with scoop
used to insert jelly-
like substance
under arm vein to
fight infection.

5"

needle
inverted

Tube of light
filaments; went up
and down.

Metallic cup,
held metallic
liquid. →
(approx. 8
oz.)

A subordinate?

John, about 15
feet away. →

Doc

A small
grey.

Beth →

Glowing mist around us all;
about one foot off floor and
six feet high.

*Figure 18. Various images that came to Beth's mind from her many
abduction experiences.*

62. Anna

While Beth was recalling her strange messages, I didn't remember much of anything. Since she was the "mother to the sisters" maybe that was appropriate. The sisters were to be kept in the dark until
the time came to do something. But I felt left out. After all, it was my

future, too, and I wanted to know what was going to happen. I don't know if I was picking up on Beth's fears, but I began to think that we were going to be moving someplace else, and soon. I had the feeling that what she was telling me was true, yet I didn't have conscious memories of anything specific, only that it felt right. I wasn't frightened, just concerned. But I needed to find out what was happening to me. At about this time we were again experimenting with our own hypnotic awareness sessions, and even though we weren't specifically looking for messages, they came floating out of my subconscious.

One of my earlier concerns had been why I was asleep when Beth experienced the missing time that started our awareness of alien abductions. One night we decided to explore those two evenings in December of 1991 to see if I had any inkling of what was going on. I did! While Beth was having her experiences, I, too, had been abducted. Nothing extraordinary (do I really mean that?) happened in these two abductions. I was given what amounted to a physical exam, had my navel punctured, had stuff injected into my body, was put in the "disinfection" showers, and had to learn about stars. I was told that all this was necessary in preparation for "the final phase."

Over several months, doing our own hypnosis and through sessions with Norris, I came to realize I had been given many messages. Most of them related to leaving this planet. Maybe this was the explanation for all my dreams of travelling and packing suitcases. I had been told several times that I must prepare to leave. I would be leaving this planet soon, yet I had a lot of things to learn first. The continued insistence on my learning about the constellations, the stars and space travel had been reinforced several times by James.

James is the equivalent of Beth's John of the security force. James looks exactly like Beth's drawings of John, but he has lighter hair, almost blond. Whenever he's around, his primary job was to intimidate me, tell me that I was holding up progress by not learning what I was supposed to. He made me feel like a small child being chastised by an adult. He once even resorted to some sort of emotion, holding me on his lap and telling me that things would be easier if I would just cooperate. He warned me again that I must not eat fat, that I had to study, that time was running out.

I did find that my inability to eat food, even with very little fat in it, increased every time James showed up. One of my ways of fighting back has been to resist learning about constellations and the stars. I bought several books, but resisted reading them. Maybe if I wasn't ready, I wouldn't have to go. Childish, but it did give me a feeling of power over them and their desires. It just feels good to be able to resist, even if I know it is probably futile.

In one of my sessions with Norris, he asked me about leaving. He distanced me from the emotions of the memories by having me describe what was happening to the ten year-old Anna.

Wednesday morning, October 6, 1993

Norris: Ask him what's going to happen later.

Anna: I'm going to go away and not come back.

Norris: Is the earth going to be destroyed?

Anna: No.

Norris: Are you going to a different dimension for further work?

Anna: Going someplace else.

Norris: Which year is this going to happen?

Anna: 1997.

Norris: Where will the little Anna be living in 1997?

Anna: Someplace else.

Norris: Someplace north?

Anna: I don't know.

Norris: What's the little Anna feeling at the moment?

Anna: Peaceful.

Norris: Has the fear abated?

Anna: Yes.

Norris: What is the look on her face? Peace?

Anna: Resignation.

[Note: The reasons for our leaving this planet have also come from my subconscious. I had called myself, and other abductees, changelings in an earlier part of this session.]

Norris: What are the extraterrestrials, the gray ones, doing? Are they helping the changelings, or are they drawing something from the changelings, are they taking something from the changelings?

Anna: They caused it.

Norris: So, the extraterrestrials caused it. Right. And they caused it thousands of years ago on the planet Earth?

Anna: Yes.

Norris:	And is it their intention that eventually the changelings will direct the Earth?
Anna:	No.
Norris:	What is the intention?
Anna:	We'll be the survivors.
Norris:	There is a holocaust coming?
Anna:	*[Deep sigh]*
Norris:	Right. When is this holocaust to come?
Anna:	It's now.
Norris:	It starts in this year?
Anna:	It's already started. It gets real bad in '97.
Norris:	What are some of the conditions that will be observed?
Anna:	Lots of winds.
Norris:	What else?
Anna:	Changes, trees get knocked down.

I saw trees broken in half—the tops missing. I also saw floods and a very dark sky, but didn't mention it because it seemed too Biblical.

Norris:	Destroyed by war?
Anna:	No.
Norris:	What do the extraterrestrials, or gray beings, gain from all this? What is their benefit?
Anna:	Some of us to start again.
Norris:	Is this plan under a divine pattern?
Anna:	They seem to know that something is going to happen.
Norris:	Is this akin to what has been talked about by Nostradamus or Revelations in the Bible, this period of chaos?
Anna:	I don't know.
Norris:	What is your role to be in this process?
Anna:	I don't get to stay. I leave.
Norris:	Where do you go? Have they told you?
Anna:	Another planet, to start again. Things are different.
Norris:	How do you feel about that?
Anna:	It's okay. It's not going to be very nice here.

I had also been shown scenes of a war at sea. Great ships were being attacked by airplanes; there was fire and smoke everywhere. I didn't know what to make of that. I'd also recalled where I was supposed to be going.

Norris: Ask them why they are genetically interbreeding with humankind.

Anna: It makes it easier to survive later, after the change.

Norris: Would you ask them if they have the right to do this?

Anna: They're doing it for our own good, otherwise we'd all be gone.

Norris: But who is to say they are not doing it for their own good? There has been much talk that they are a dying species. Is there truth to that?

Anna: I don't know. I get mixed messages. They live for such a long time.

Norris: Ask to know what is their home planet.

Anna: Arcturus?

Norris: Arcturus. Is that the only one?

Anna: Well, that's the original one. There's lots more now.

Norris: That is the original planet, Arcturus. What are some of the other planets?

Anna: Beni, Bene, Beneton?

Norris: Beneton, good. They'll come to you easily.

Anna: Pleura, Pleurad, Pleuraine.

Norris: Plurad, Pluerane. Are there more that you can recall?

Anna: There's the red planet that I'm going to, but I don't know it's name.

[Note: A later memory indicated that it may be named Feyda.]

Norris: So you're going to the red planet.

End Transcript

Do I believe any of this? I don't know if I do, it's just there. I have accepted it as just another piece of the puzzle. The gray ones lie for their own benefit, and maybe ours as well. It's impossible to sort out which are lies and what may hint at truth. Or was my mind just giving me an explanation for what had been happening? It didn't matter. It seemed real.

The room I visit with the large viewscreen showing stars also has another purpose. I have been taught to operate the controls of the ship. There don't seem to be many levers to move, buttons to push, or dials to be read—nothing like I see on the starship Enterprise. I do move my arms and my body to control the movements of the ship, but it all seems to be body-English. I don't feel that I am connected to the ship's guidance system by wires or other mechanical means; it's as if my whole body becomes a part of the guidance system. I feel the ship change direction or altitude based on my motions. I don't think one should be able to feel that in space. After all, it's not like a car careening around corners in Earth's gravity. But that's what it feels like. I get to practice that a lot. I'm not very good at keeping the ship level and going in one direction; I do a lot of yawing.

Is my next job to be that of a navigator or pilot? If so, I need new job skills.

63. Beth

Norris was a real sweetheart. His hypnosis techniques were relaxing and provided me with some degree of temporary acceptance. Temporary was the operative word, though. I had learned not to expect too much from hypnosis; I had enough conscious memories to last a lifetime! And the clarity with which I relived these experiences under hypnosis often caused me considerable anxiety. The benefits, I thought, were highly overrated. Yes, it was true that after one of these sessions it was easier to talk about the experience I had just recounted, but since so much more was revealed this way, the trauma associated with the memory was also more intense. In short, I was beginning to view hypnotically retrieved memories as just more fuel for the fire.

Hypnosis had certainly been useful in uncovering lost time, and as a means of implanting post-hypnotic suggestions to help the subject deal with these emerging memories it had proved invaluable. The procedure had demonstrated its worth over and over again, so my disillusionment had little to do with the method itself. We had taken pains to choose hypnotists who were more concerned with our emotional well-being than with obtaining firsthand information to satisfy curiosity, so it was not the practitioner who had fallen down on the job. It was me. Whatever gains had been made through the practice of hypnosis were duly noted and appreciated, but it was now time to move on.

Norris, like so many others, was curious about our experiences and the beings we described. A self-proclaimed New Ager, he wondered if the grays might be of a more ethereal nature, their origins predating the emergence of humankind, even before life itself evolved, and that they may have been instrumental in our development all along. Since we had no evidence to contradict these theories, we were in no position to argue the point. Even so, I was not yet ready to accept Norris's explanation over our own—that these creatures were not of this world and that their agenda did not include saving us from ourselves. I hadn't gone so far as confining them to the *bad guys* camp, but neither could I credit them as *good guys*.

The abductions were frequent, the sporadic and fragmented memories powerful and unnerving, the messages insistent and disquieting. Anna and I believed, by this time, that the aliens had no intention of telling us any more than we already knew about their motives and their intentions for us. We had been forced to assume a great deal, lacking sufficient information to make even a rudimentary guess. Some of our assumptions were, to put it mildly, outlandish! It appeared, from our remembered experiences since childhood, that we had been influenced to develop in certain ways with specific interests. Anna, for example, developed a desire to study marine biology. I had such a love of animals that my parents were driven to distraction trying to keep me from bringing home every stray animal and wild beast I found.

The most intriguing influence brought to bear on experiencers seems to be the deliberate pairing of certain individuals since early childhood. How this is managed we can't even imagine, but certainly Anna and I were manipulated to assure our meeting again as adults. To what ultimate purpose? Are these creatures so powerful they can successfully direct our course, confident we will blindly follow their lead?

These incidents, taken individually, could be attributed to coincidence, but when our memories of interaction with the greys during childhood are taken into consideration, they become prophetic. Frequently during an abduction I am shown a variety of animals, both domestic and wild and in all stages of development, then instructed to study them. My interest in horses has been encouraged at different times. I was even told by Doc that I could talk to them, if I'd only learn to listen! It would also seem that I have been encouraged to use my artistic talents to reproduce in art form nearly every species of mammal in existence!

Messages that hint at some future catastrophe are commonly received by experiencers [both abductees and contactees alike). Instructions are given to experiencers that they congregate at specific "meeting" places where, presumably, they will be collected one last time, but will never return to earth. As difficult as it is to accept these messages, it is equally difficult to ignore them. The messages, like the other influences, have a strong hold on the experiencer's psyche and daily routine—never very far from mind.

Are these subtle influences a means to an end? Who knows! But one can't help wondering. And no hypnotically retrieved memories have answered that question, or so many others.

chapter 12

Where Does It All End?

64. Beth

The whole event was clear, from start to finish, as easily remembered as what I'd had for dinner the night before. The only time I had come close to such a complete recall was my abduction from the cabin over a year ago—and that had a few holes in it.

I had woken up at 2:35 A.M. to find my bedside lamp on. I knew I had turned it off before going to sleep. Fully alert, I reached over to turn the light off and got a painful shock. Surprised, and a little nervous, I sat up in bed and looked around the room. Everything was normal—no shadows lurking in dark corners, no electrical discharges, no indication that anything was about to happen.

Suddenly the lamp snapped off of its own accord, but the room was not thrust into darkness. It was filled with light! I heard a popping sound, then saw a hazy film cover the window that faced the front of the house. The haze dissipated and two of the little greys (the escorts) appeared in its place. They floated toward the bed in unison like toy soldiers manipulated by an invisible giant. Stopping just inches from the bed, they waited, as if I should know what to do. An indeterminate time passed as I stared at them and they stared at me, until finally I felt myself rising from the bed. I was turned in midair, then floated *through* the closed window and up into a black hole in the sky.

The blackness swallowed me, then closed itself off, and I was deposited in a brightly lit oval-shaped room of unknown proportions. I knew where I was and what was expected of me, yet I couldn't remember when I had last been in a similar place. I was led by other escorts to another room where I was instructed to lie down on a padded bench. I thought about refusing, then remembered that they could easily make me comply, so I did as told. The bench felt soft and cool, but not cold, and I realized my nightshirt had been removed. When had that happened?

I was rolled onto my left side, my right leg bent slightly at the knee and allowed to overlap the left, presumably for stability so that I wouldn't roll right off the bench. My left arm supported my head, the other was draped across my breast. It was not an uncomfortable position and under other circumstances I might have dropped off to sleep.

I sensed movement, but had trouble focusing and couldn't lift my head to look around. I wasn't particularly afraid, just annoyed. If they left me alone, I could catch a few winks! Then I saw something out of the side of my eye: A large oval-shaped object was lowered from above, stopping about a foot from me. Initially I couldn't tell how long it was, but the lower end went as far as my knees. It looked like a giant elongated eye or camera lens. The "iris" was round and clear, like glass, the rest dark, almost black. Was this some kind of x-ray?

Without warning, the machine came to life, it's rim emitting a reddish pulsing glow. I felt heat, but it wasn't uncomfortable. Then the round center flashed a bright white. Now the heat was more noticeable, and I instinctively "called" for Doc. If there was going to be pain, Doc could take the pain away. But he didn't come. The shadows bustled about as before, but none came to comfort me. The machine started to move, scanning up and down my side, the pulsing red glow and brighter white light beginning to burn my skin. (See Figure 19.) I felt like a document trapped inside a photocopier! The burning intensified and I screamed silently for Doc to come and help me. Why was he letting me suffer? Were these rogue aliens experimenting without Doc's knowledge or permission?

Finally the white light blinked out, and the machine stalled in its original position with its pulsing red rim fading more slowly. My skin felt as if it had been fried to a crisp; it tingled as the intense heat was replaced by cool air. I forced myself to breathe evenly and slowly, filling my lungs. I felt wetness on my outstretched arm and knew it was tears. I began to relax, convinced the ordeal was over, when I felt my left leg being pulled out behind me. There was a sudden prick—not painful, by comparison—in the back of my thigh, but I hardly noticed, so eager was I to be released from my tormenters.

The leg was repositioned and I prepared myself for freedom, but the "scanner" was again activated. It resumed its sweep of my side, the burning so agonizing that I screamed. I knew I hadn't screamed aloud, that I had only screamed in my head, but they had always heard me before. Surely someone, even a lowly tech, would not let me die there!

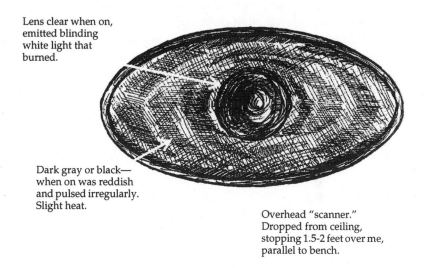

Lens clear when on, emitted blinding white light that burned.

Dark gray or black— when on was reddish and pulsed irregularly. Slight heat.

Overhead "scanner." Dropped from ceiling, stopping 1.5-2 feet over me, parallel to bench.

Figure 19. Scanning "light" that burned Beth's skin.

To my great relief, I saw a hand descend on my left shoulder, just under my chin. Somehow I knew it wasn't Doc, but I expected the pain to ebb. It may have, but being temporarily distracted by this first really close look at one of their hands, the pain may have taken a back seat. I studied the hand closely, noting every detail. I could draw that hand fairly accurately—if I remembered I had seen it! (See Figure 20.)

The "scanner" shut down once again and I held my breath. Was this *finally* the end? The entire mechanism rose back up to the ceiling and stayed there, just out of my view. The hand on my shoulder was withdrawn and I was told I could go.

I was brought back and placed in the center of my bed, my nightshirt crumpled under me. Apparently it had been removed before I was taken. Not bothering to put it on, I laid down and went directly to sleep.

In the morning I awoke with the memory intact. There was blood on the sheet and I knew why. Examining the back of my left thigh, I found two puncture marks; the right side of my body—from shoulder to thigh—was inflamed and tender. I was thrilled that I had remembered what went on, and how these injuries had happened, but frightened by them as well. Maybe it was better not remembering!

Over the next two weeks, all the skin on my right side peeled off—twice—leaving raw flesh that was dangerously exposed to infection. Fortunately, infection didn't set in, but it was some time before the skin layers grew back.

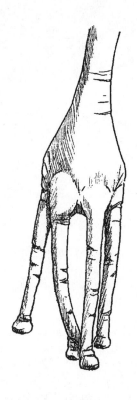

Figure 20. "Tech's" hand.

Was this how it was going to be from now on? I wasn't sure I could handle having conscious memories. I thought I could...before. Why was I remembering now? What had triggered it? Perhaps the amnesia was self-induced and had nothing to do with *them*. If that was true, then my mind had been trying to protect me against the trauma of these experiences by repressing them. Didn't I need that protection anymore?

If the aliens were responsible for blocking these memories, then what had changed? Why did they no longer feel the need to block them?

What if this one was just a quirk, like the other one last year? That was certainly possible. Maybe things would go back to the way they were; I won't remember more than a few pieces, if anything, and life with Papa Doc will go on as before.

I had decided that full conscious recall was the exception and not the rule—until a few weeks later. After a much less traumatic experience, I recalled everything right away, down to the smallest detail. I didn't feel any more able to cope with these instant playbacks than I had before, yet there was no choice now. Something had happened that changed the rules and regs, and as always, I had no say in the decision.

65. Anna

Telling my father was one of the hardest things I've done in the past several years.

It seemed like I'd spent a lifetime avoiding talking to him about anything that mattered to me. Lately, I'd even avoided talking to him about anything. I felt so ashamed that I had believed he had raped me, and now I knew that wasn't true. I had a lot of years of hate, anger and frustration to atone for. I wasn't ready.

I answered the phone a few months ago and it was Dad calling to see how we were. There was no one else in the house, so I couldn't do my usual and speak briefly to him and then give the phone to my brother or sister. We had a pleasant chat, about nothing in particular; everyone was fine, nothing new on the farm, he was healthy and happy. I was my usual distant self. When we hung up, I broke into tears. Why was I doing this to him, to myself?

I immediately called him back and tearfully apologized for how I'd been treating him all these years. He didn't understand; he was confused. So I told him. Not all the gory details, just that Beth and I had been having some strange experiences and memories, and that we were convinced that we were being abducted by aliens. I explained our lifelong association, our memories of being brought together in strange places, by strange beings. He had no knowledge of anything odd going on in my life, but didn't doubt that what I told him was true. He was sympathetic. He understood!

· My father has always been an avid reader, and lately, much more open minded than I remembered him as I was growing up. He said he had read about UFOs, and knew of the possibility of abductions. He had always believed that there are stranger things around than we can even begin to understand, and this new knowledge fit right into that category. He asked how he could help us. I didn't know, I was too overwhelmed.

I mentioned that I also thought that Mother had been involved. I didn't dare mention my brother Rick's involvement. Rick hadn't fully committed himself to that idea yet, and I didn't want to broach the subject. I sent my father a list of books to read so that he would have a better idea of what we were experiencing. I couldn't tell him myself. The years of misunderstandings had distanced us too much for me to talk freely about things that meant so much to me. But it was a good start. The next time he visited I would be more prepared to talk—so might he.

In the next few weeks, I felt more able to examine my family relationships and the devastation that the alien intrusions had caused. I was very distanced from my father, and somewhat less so from my two sisters, yet always felt very close to my mother and brother. For me, the family seemed, now, to have been split into those experiencing abductions and those not so encumbered. But how could I be sure? All I had was intuition to go on. I had partial memories of a few incidents, but nothing really specific, except a strong feeling that my mother and Rick had shared in these experiences.

My need to know prompted me to ask Beth to use hypnosis to help me. The strategy we decided upon for the session was simple: Beth would ask me to go back to a time when my mother and I were alone, and something unusual happened. If that didn't bring about any abduction memories, we'd try the same strategy using my brother's name. Yes, we knew that Beth would be leading me; we were trying to find abduction memories, but thought that the open-ended question would work. If it didn't, I'd be no more in the dark than I was. If it did work, then I might be able to make more sense out my childhood resentments.

After I was comfortable and hypnotised, Beth asked me about my mother. My first memory was of being in the kitchen, cooking dinner. My mother and I spent a lot of time together cooking as I was growing up. She had a flair for gourmet cooking, and I developed my love of food and cooking from her. I must have been in high school at the time. The unusual incident occurred while I was making mashed potatoes. I had added too much water and they turned to instant mashed potatoes, and then burned! So much for an abduction. We moved from that remembered incident and tried again.

The next memory was that of our being on the beach together, after school one evening. I must have been in junior high school at the time. It was not unusual for the whole family to go to the beach after school, have a quick swim, and then have a picnic supper on the beach. It was unusual that just my mother and I had gone. Everyone else must have been busy doing other things. A storm came up, wind and rain forcing all the other swimmers from the ocean. My mother and I continued walking along the beach and noticed a fog bank coming in off the ocean.

The next memory I had is of being enveloped by the fog and then being on a craft. Two gray beings took my mother away, telling me that she had other duties to perform. I was led into a room with a large viewscreen and seated at a "desk." The viewscreen showed an underwater scene. I don't know if the ship was literally under water, or if it was a projection. I remember being told, "You must study the creatures of the water." Nothing else came from that memory.

While in junior high school, I was awarded first prize in the Science Fair for my extensive collection, knowledge and management of tropical fish. I had fish tanks all over the house and even dabbled in salt water aquariums as I got older. In college, my major was Marine Biology. I didn't graduate with a degree in it because my major professor and I just didn't see eye to eye on too many things. My father taught me to scuba dive when I was about fourteen, and I obtained my certification in 1970, when it became difficult to purchase air without a card. I have spent over thirty years scuba diving, and still find I am as comfortable under water as I am on the surface. Did the aliens direct my love of the water? Even if that's true, it's been one of the most pleasurable aspects of my life.

Beth admitted to me after the session that she had also seen my mother aboard a craft, many years ago. At that time she had recognized the woman, but didn't know she was my mother. She associated the woman she saw on the table as A.J.'s mom. I'm glad she hadn't told me about that before I had retrieved my own memories.

Since these memories were not that traumatic, and they were brief, Beth decided to continue. She asked me to go to a time when my brother and I had been alone and something unusual happened. My first memory was from the high school years when we were fishing together. We were out by the jetties that flanked the entrance to the harbor. A large sailboat was negotiating the mouth of the harbor, and ran into the rocks! It was not badly damaged, and sailed off with only a small hole in its hull. Yep, that was something unusual, and it didn't involve the greys. She tried again.

I saw Rick and I again on a fishing trip, but this time I had caught a ladyfish. These fish were hard to catch, and quite rare. Beth tried again.

This time I went far back to a time when I was about five years old and my brother was three. It was winter and we were in Wisconsin, playing outside with our two cousins. Rick was on a sled, since the snow was too deep for him to walk easily and we wanted to explore the woods. While we were in the woods, we were approached by two short gray beings. Rick and I (and the sled) went with them, my cousins stayed behind. We ended up in a small craft where we were separated. I was put on a table and had an incision made from my breastbone to my groin. I wasn't scared; I was fascinated. It didn't hurt, there was no blood. The aliens were poking around in my innards as I watched, lifting each organ and examining it. They said they were checking that everything was all right. After they finished, they brought a wand down from the ceiling that emitted light. They ran the wand over the incision and it closed back up again. They then rubbed a blue gel on my stomach. I have no scar. There is no indication that this ever happened to me.

I was then returned to my brother, who was waiting for me in the room where we had first entered the craft. He was sitting in an alcove, levitating himself up and down, laughing. We afterward were returned to the woods where my cousins were waiting. We had an argument over who was lost. They had been searching for us. I told them they had been lost and that we were here all the time. As it was getting late, we returned to the house where we all got chewed out for being outside so long. We each blamed the other pair for getting lost. I was particularly singled out because Rick's boot was missing. No, I couldn't explain where it was. I had taken good care of him. He must have lost it by himself, I believed. I wonder now, if the aliens might have a storeroom full of clothing they have forgotten to return, or if some other child ended up with three boots.

My intuition seemed to be correct again. My brother and mother had been involved. I also felt relieved that our hypnosis techniques had not let us down. It wasn't that they had uncovered abduction memories. It was that in several instances, they had uncovered other memories that were unusual to me at the time, but didn't involve abductions. Maybe I wasn't making all this stuff up.

66. Beth

It felt good to be painting again. I had been away from it for too long and was probably rusty, but the therapeutic value alone couldn't be matched by anything, in my opinion. Sitting there before the blank canvas, my mind completely absorbed by the images I hoped to recreate, the outside world didn't exist. Doc didn't exist, the aliens didn't exist, UFOs didn't exist—my whole world consisted of a sixteen by twenty-inch stretched canvas perched on an easel. Tubes of acrylic paint were laid out on a nearby table within easy reach. An assortment of pigments had been selected and squeezed out on a palette, ready and waiting. The bright dabs of color looked like tropical fish suspended in a strong head-current. They were glossy and fresh, more so than I was accustomed to. I had learned to paint with oils and had never experimented with acrylics before, but oil paint took too long to dry and I wanted instant gratification, paintings I could finish and move the same day.

There was no demand for speed. I didn't have an order that needed to be filled by a set date, no reason to complete a painting by the end of the day, yet I believed it was time for a change. Learning a new technique using acrylics would be a challenge to my talents, and I was up for it!

The first couple of paintings were rough, and I wasn't anywhere near satisfied with them. There was much room for improvement, and I needed lots of practice learning to use this new medium. I had heard somewhere years before that an artist who is satisfied with his work is not admiring talent, but imagination. If that was true, I was either a promising artist or totally unimaginative!

On my next day off, I hurried to my studio eager to get started on another painting. This one was to be a jaguar. I wasn't confident enough of my knowledge of the animal to do it justice on canvas, though, so Anna and I had looked around the day before for a clear photo of one in the wild. Eventually we located a fairly good photo and I took it to the studio with me. Working continuously through the day, I finally finished the painting and presented it proudly to the girls and their brother, trying not to look too pleased with myself. Actually, I thought it turned out quite good—much better than I had expected—but I had to keep reminding myself of that quote. Maybe it wasn't as good as I thought!

I was encouraged nonetheless, setting aside time every day to paint. Within a week, I had completed five more paintings, all portraits of animals in their natural surroundings. This preoccupation with wild animals was not chance: I had been "told" to concentrate on endangered species, and that each painting was to be given away as soon as it was completed. With my love of animals, painting them was not exactly a chore! And I didn't mind giving them away, either. I wasn't doing them in order to make money—though in previous years I had supplemented my income by doing just that.

The work continued at a furious pace; paintings were produced one after the other and given away almost before the paint was dry.

I had begun work on a Siberian tiger, an animal I considered worthy of my best efforts—but it was giving me trouble. I knew how I wanted it portrayed and had numerous examples to refer to from a half-dozen *National Geographic* magazines, yet my attempts to properly recreate this regal cat just wouldn't come together on canvas! Needing a break anyway, I got up and stretched, standing back from the canvas for a better perspective. I hadn't had this much trouble with any of the others and couldn't figure out why it was such a problem. Something in the animal's coat...its texture and softness....

I heard a dull pop, like the sound a balloon might make if it was ruptured under water. There before me, as real as anything I had ever seen before, was a live, breathing tiger! I jumped back, not wanting to believe this could be real. It had to be my imagination! I had been working so hard to get it right....

Then I heard a familiar voice in my head: *"Touch the animal, feel its coat. You must understand how it is made and how it moves, then you will be able to reproduce it. Feel its fur...touch it and study it."*

Reaching over obediently, I stroked the cat's fur. It felt warm and dense, but surprisingly, not as soft as I had expected. It wasn't like a domestic cat's fur—not downy or silky to the touch—more like what I thought a bear's coat would feel like. The tiger stood absolutely still, yet I could feel its breath; its sides moved in rhythm to its breathing, its ears flicked back and forth. It *was* really alive! I couldn't believe I was petting a wild tiger in my studio! I wasn't even afraid! Cautious, maybe, and respectful.... How did it get there?

I straightened slowly, not wanting to startle the cat. In the blink of an eye, it was gone! Hesitantly, I moved over to where it had stood only a second before. There was nothing there but empty space—no footprints, no wild animal smell, nothing but bare tile floor. Had I imagined the whole thing? I must have been working too hard, spending too many hours in seclusion doing one painting after another. I sniffed my fingers; surely the animal would have left its scent on my hands when I stroked it, but they smelled only of paint. How odd.

I finally sat back down and got to work. Now the painting went along smoothly, as I altered my brush strokes to reflect the tiger's rougher, more dense coat. The painting was finished in two hours, and I was happy with it. It looked real to me. Maybe that was because it was real in my memory.

One after the other, the paintings poured forth like a rain-swollen stream down a mountainside. There seemed to be no end to my stamina or enthusiasm! But the subjects of my paintings did change. After a particularly vivid dream where I was told by Doc that I should document different ethnic groups, I abandoned my noble animals in favor of human portraits; Indians, Peruvians, Japanese, Chinese, Russian Cossacks and more—all in native costumes. The research involved in this new project was much more difficult since my knowledge of some of these cultures and their dress was extremely limited. But the painting progressed—slowly and deliberately.

The painting continues, and I look forward to the next project, no matter how or from where the inspiration comes.

67. Anna

I was now pretty comfortable relying on my intuition. It had proved valid in many instances. But now it was telling me something different. I felt that I could heal people.

I don't know where the idea came from, it was just there one day. I had not accepted my abduction experiences at that time, so I didn't consider that I might have been abducted and the thought implanted in my brain.

I first started having odd feelings about people's health on the trip to New York when we visited Budd Hopkins. We were sitting in a restaurant eating lunch when I glanced at a person at a nearby table, and I knew he had AIDS. No, I didn't ask him; I just felt uncomfortable with that insight.

My next intuitive leap came as I was talking to a friend at the farm. She had been having problems with endometriosis and was considering having a hysterectomy. She was going to a doctor the next day to have a full checkup. I asked her if she was going to have a mammogram. I had a precognitive flash, and I saw her without a right breast. Where the breast should have been was a dark lump. I also saw a large mass around her left ovary. The results of her mammogram showed a suspicious area in the right breast, but no tumor.

During her hysterectomy, the doctor removed a large, grapefruit-sized mass from the left side of her uterine cavity.

One night I had a dream about a friend I had met at Budd's support group meeting while we were in New York. I saw her on a bed and the side of her neck was swollen. When I woke up I felt I had to call her and warn her about her thyroid gland, but I felt foolish doing so. I didn't call. A few months later we met and I mentioned that she probably should have her thyroid looked at. I felt that there was something wrong with it. No problem. She'd already had it diagnosed, in the last few months and was now on medication.

In a few cases, at least, my diagnostic talents had proved out, but what about the healing? I first tried it out on a friend's daughter. She had been diagnosed with cancer and was getting ready to undergo more tests and start therapy. I envisioned macrophages eating away at her cancer for several weeks as I would drop off to sleep. A few months later, my friend told me that his daughter had been given a clean bill of health. The cancer disappeared, or had been misdiagnosed. We didn't care; it was gone. The image of Pac-Man eating up cancer cells was ludicrous, but maybe it worked.

I next decided to work on one of our horses. He had a respiratory infection several years ago and this had developed into a continuous cough that made it difficult for him to be ridden comfortably. We had tried several medications over the years, but his symptoms only disappeared for a short time. His breathing had become so bad that he had developed a heave line in his abdomen, an extra band of muscle used to exhale. Our veterinarian had told us that his disease was progressive, and incurable.

As I concentrated on him, I tried to envision his lungs and his respiratory tract. I saw them inflamed and bright red in color. I tried to change the color to a healing electric blue. I worked on him for several weeks, finally also enveloping him in a protective purple light. It just seemed the right thing to do. I tried the "laying on of hands" that I had seen done by healers on TV. He wouldn't let me! I guess that wasn't to be my way of healing. One day as I was feeding the horses, I took a close look at him. He no longer had the heave line, and Beth had reported that he almost never coughed any more. Maybe it worked. He's back in work now and seems healthy. I still concentrate on him every week or so, just to keep him healthy.

Since that time I have worked on other people's, and horses', illnesses with minimal success. I feel that what I do, if anything, takes a long concerted effort. It's not the overnight healing that others have reported. Does it really work or am I fantasizing? I don't know. But it doesn't take *too* much effort and I usually try and do it as I am drifting off to sleep at night. The only physical effect I have noted is that my body temperature seems to rise as I concentrate. Too bad I haven't been able to make it work on myself!

68. Beth

I had been taking evening classes through our local community college for three months, setting aside every Thursday night in hopes of completing an intense course in equine management. Firmly believing that one can never know enough about one's job, those four hours one day a week could make the difference between a satisfactory job performance and an educated, enthusiastic one.

In April of 1993 as I was leaving my Thursday night class, I noted the time: It was just after 9:00 P.M., earlier than usual. I spent a few minutes in the parking lot talking with the professor, then excused myself and left. I was tired, but expected to be home before 10:00. I could relax then.

I took the usual route home, a country road that wound past horse farms and across wooden bridges. It may not have been the fastest way for some, but I knew the roads well and could drive along at a pretty good clip. Besides, there were no stop lights to worry about and no traffic. Driving past a turnoff to an historic mill, I noticed a pickup truck parked on the gravel shoulder. There were homes scattered through the hills nearby and the river running past the mill was a popular fishing site for those residents, so it was not uncommon for vehicles to be parked on the shoulder during good fishing weather. But I had never seen one parked there at night. Deciding it wasn't worth worrying about, I drove on.

As I turned a bend and began the climb out of the river valley, I saw lights reflected in the rearview mirror. Assuming someone had pulled out of the mill area behind me, I took no special notice. Suddenly I was blinded by lights glaring through the back window! I tipped my rearview mirror so that the light would be reflected back to the inconsiderate driver, but it didn't help much; the lights apparently were on high beam. Slowing down, hoping the idiot would pass, I watched through my side mirror and noticed that the vehicle was a pickup, but I couldn't see the driver or if there were any passengers. It looked like the same truck I had seen parked on the shoulder near the mill.

When the driver didn't take the opportunity to pass, I picked up speed, expecting to outdistance him easily. (I didn't know for certain the driver was a man, but somehow I couldn't believe a woman would have any reason to run down another car on a back country road—especially if given room to pass.) The truck stayed with me, close enough to my bumper to hitch a free ride, as I accelerated to over sixty. In good conditions, I had been known to drive fifty or fifty-five on short stretches of that same road, but having a truck with its high beams bearing down on me at those speeds was treacherous.

As I approached two wooden bridges, the first at the base of a steep hill with a hairpin turn to the left, the other on the far side of that same hill, I was forced to slow down or risk losing control and flying off the road. After negotiating the hill and crossing the last bridge, I again slowed down for the truck to pass. As before, he refused to, matching my speed and staying close behind. We weren't far from a crossroad where I planned to pick up another route, so I drove on ahead, sure this maniac would eventually turn off.

Arriving at the crossroad, I turned right then had to stop at the railroad crossing before going on. (This same set of tracks had to be crossed twice before reaching the main highway; the second crossing was less than two miles away.) I checked the rearview mirror and saw that the truck had stayed with me. There was nothing I could do about it! I certainly wasn't going to pull over. There was no telling what kind of person this was or what he might be capable of.

I intended to turn left after crossing the tracks, but looked right first to be sure there was no oncoming traffic. I saw flashing lights and road flares in that direction, indicating a possible accident or breakdown, but since I wasn't going that way I concentrated on just getting to a main road in one piece! I checked the time again: 9:25. With the road blocked by flares, no traffic could get through, so without turning on my indicator, I hurried across the tracks and turned left, spewing gravel in my wake. This route was fairly straight, but hilly, and I knew I could safely travel at higher speeds—so I floored it. But the truck caught up, riding my bumper just as before! When I reached the next crossing, barely slowing down, I realized the truck had disappeared. He must have turned off somewhere, I thought, relieved to be rid of him.

Once over the tracks I slowed down considerably. The road ahead was straight and flat, dead-ending at a two-lane state highway. Here I would turn left and follow it for about three miles to one more turnoff—then home. Nearing the intersection I saw a row of lights directly across the highway. It looked like a circus! There was a farm supply store in that same area, but it had closed hours before so there shouldn't have been any lights on around the store or parking lot, which abutted the highway. As I got closer, I realized they were headlights from vehicles parked facing the roadway—and directly into my eyes! Great, I fumed. Another exciting adventure. Even though I had probably made record time, I felt as if I'd been driving all night. I didn't feel up to yet another diversion.

I braked at the stop sign. Another accident? There were flares in both lanes to my left, and more to my right. A roadblock? I recognized a couple of county police cars, one or two state troopers, and—oddly—some military jeeps and trucks with camouflaged canvas sides! Had some dangerous criminal escaped? I immediately thought about the truck that had tailed me and then suddenly vanished. I might count myself extremely lucky before this night was over.

A man in plain clothes signaled for me to pull out and stop in front of the flares. I rolled the window down a few inches, just enough to talk through, and asked what was going on.

"May I see your operator's license?" he said in response.

"Sure," I replied, digging through my purse hoping to locate it by feel. My dome light had not worked since the gray shits had fused the wires, so when I was unsuccessful, I asked the officer if he would please shine his flashlight into the car. That helped, and I found it easily, holding it up to the window for him to see.

Evidently that was not satisfactory. "Please hand it to me," he ordered. After inspecting it and finding it in order, he passed it back through the narrow crack, thanking me politely.

Assuming I could continue on my way, I started to put the car in gear when I heard him ask me to step out of the car.

"Why?" I demanded. "What's going on?"

"Please step out of your vehicle," he repeated more forcefully.

This person was not wearing a uniform, and although I could see a county sheriff's deputy guarding the flares, I had not been shown any identification from this man and wasn't about to get out without a reason. So I asked him to show me his ID.

"That's not necessary," he announced. "Step out of your vehicle!"

Now I was afraid. Something was wrong with this whole scenario. I hadn't done anything illegal, my license was obviously in order, and this man refused to identify himself or tell me why I was being ordered out of my car. So I refused—and held my breath. The man backed up a few steps, signaled to the officer ahead to clear the flares from the lane, and I was ushered through, free to go home at last!

I thought....

As I passed through the maze of road flares and picked up speed, I spotted the pickup on a side road. I was only a few miles from home and had no intention of letting anything else interfere. As expected, the truck pulled in behind (though he had plenty of time to pull onto the road ahead of me). I had reached fifty-five, but the truck was gaining fast and was soon hugging my bumper again. When I reached the final turnoff, I swerved recklessly across three lanes and gunned it down the side street. The truck stayed close all the way to the farm's entrance, but went on its way after I turned into the driveway.

Home never looked so good! I parked the car across from the house and reached for my books, but they weren't on the passenger seat where I'd put them after class. I felt around on the floor and finally found them. I crawled out of the car, locking the door, then remembered that the passenger side was unlocked and had to go around and lock it. Every little thing seemed to take extra effort and concentration! As exhausted as I was, though, I couldn't wait to tell Anna about my harrowing trip—but she was already in bed. It was still early, I thought, even for Anna. But when I checked the time. It was almost 11:00!

I'd been abducted! I couldn't blame this one on highway hypnosis; there hadn't been a moment when I wasn't fully alert—and worried. I certainly hadn't been lulled into dreamland. But why didn't I remember what happened? Lately, I'd had conscious recall when an abduction had taken place. I had nothing this time. Nothing at all!

Dumping my books on the office desk, I struggled painfully out of my jacket (my arms were sore and stiff), finding that the right pocket had been ripped somehow. Oh well, I could sew it up tomorrow. I decided to call it a night. Maybe it would all come back to me in the morning.

Morning came, but memories of being abducted didn't. I remembered being hounded by the truck, seeing the roadblocks, stopping and showing my license, refusing to get out of my car, being ushered through the roadblock, picking up the truck again, then leading him home. Nothing else. If I had been abducted, I reasoned, it had to have happened sometime between the first rail crossing and the second, when I realized the truck was no longer behind me.

I wondered if the authorities knew what had happened during this time…. Was that why so many of them were at the roadblock? Was that why there was no traffic? Had I been deliberately isolated so that they could check me out afterwards without drawing attention from passers-by? That was crazy! All those vehicles and bright lights would have been seen for miles! Needing some explanation—

or lacking that, some proof that what I *did* remember of that night was accurate—I called the county police and asked if they had a roadblock there the night before. They denied it, saying I should call the state police; it had probably been them. I did as suggested, but got the same negative reply. Frustrated, I asked Bob Huff the next day if he would please check out that area of the highway to see if there were any evidence of a roadblock. He had better luck, finding signs of there having been a number of flares in both the north and southbound lanes exactly where I remembered the roadblocks to have been.

So there *had* been a roadblock. I hadn't imagined that part. I felt better—sort of—but that still didn't answer my questions about the missing time.

Weeks passed and no new memories surfaced. This abduction—if there had, in fact, been one—was more frustrating than many of the others by my inability to remember it! I began reexamining the whole evening, step by step, looking for any previously overlooked piece that couldn't be rationally explained. The only things I could come up with were the torn pocket on my jacket and the soreness in both my arms. But the jacket could have been torn before and I just hadn't noticed it; my arms might have been stiff from tension. The books strewn on the floor of the car I didn't classify as a mystery; with all the twists and turns I'd made trying to outrun the truck, it would have been a miracle had they stayed on the seat!

I was at a point where retrieving these memories, if there were any, would mean turning to hypnosis. Yet that worried me, too. On other occasions I had the benefit of some memory, but this time I had nothing. Would my mind fill in the blank with fantasy? There was only one way to find out. But we were a long way from Budd, and my seeing Dr. Ruxer was financially impossible at that time. There was no one else who could help me, and I had to get to the bottom of this enigma!

But there was someone else I could turn to....

Anna and I had, through hours of practice and study of hypnosis techniques (as well as our own experience as subjects), begun to feel confident hypnotizing each other. Although we sometimes had difficulty with the deeper states of relaxation to access repressed memories from way back, we were quite successful retrieving more recent memories. My latest episode of missing time was only a few weeks before; if we were going to explore this event and hope to retrieve anything reliable, we would have to do it soon.

On a Monday evening we decided to try it, recording the session so that we could refer to it later for clarification or as a tool for comparison should any conscious memories unfold. Anna began the session with the customary relaxation exercises. This took a little longer than usual as I was trying hard not to go into it with any preconceived notions or expectations that an abduction had occurred. Once fully relaxed, Anna took me back to the night in question, just as I'd left class:

Anna: Where are you now?

Beth: Talking to Dr. Keller [the professor]. We're outside. She wants to go get something to eat, but I just want to go home. I'm tired.

Anna: Okay. So you start home. Which way do you go?

Beth: You know, the back way, through town and left on Route 643, or whatever that route is.

Anna: Okay. Is there a lot of traffic?

Beth: No, there hardly ever is. I just go—

Anna: So you're driving down that road, and there's no traffic, right?

Beth: Right.

Anna: Now at some point we know you saw a car parked on the side of the road. I want you to go to that point where you see this car, or whatever, parked on the side of the road. When I count to three, you'll be at that point where you see this vehicle parked. One...two...three.

Beth: It's parked there on the shoulder, on the left side of the bridge going into the mill.

Anna: And what do you do?

Beth: Nothing. I just drive by it. Couldn't see in the truck, if there was someone in the truck. Well....

Anna: Does the truck pull out, or is it just parked there?

Beth: I don't know. I mean, I saw lights, I think.

Anna: Did you continue driving?

Beth: Yeah. Where the road goes up the hill, out of the river valley—that real sharp turn. And I—sort of like a reflection in the mirror.

Anna: What is it a reflection of?

Beth: Just like headlights, or something, coming around the corner behind me.

Anna: Okay. Do you drive on to the village? [The small crossroads just before the first railroad track crossing.]

Beth:	No.... Up over the hill, car lights behind me, getting too close, coming up real fast.
Anna:	Real fast. And what do you do?
Beth:	Speed up! Try and keep ahead.
Anna:	Uh-huh....
Beth:	I wasn't going very fast. I should go faster!
Anna:	Did you go faster?
Beth:	I can go the speed limit.... I think. [Laughs]
Anna:	Is the vehicle still behind you?
Beth:	Oh, yeah! It's right on top of me! If I open the trunk, he can climb in.
Anna:	Oh, okay.
Beth:	It's getting really dangerous.... It's not good to do this. [*Whispers anxiously:*] Oh...I can't pull over! I'm going to slow down.... Maybe he'll just pass me.
Anna:	Does he pass you?
Beth:	No! I can lose him at the bridges. I can do those bridges!
Anna:	And do you lose him at the bridge?
Beth:	No...
Anna:	No. He's still behind you.
Beth:	Yep. I come off the first bridge and go over the hill, make that corner really fast—too fast.
Anna:	[*After a long pause:*] Where are you now?
Beth:	Trying to let him pass.... He won't pass. We're coming up on tracks. Maybe he'll turn off.
Anna:	And does he turn off?
Beth:	No! He pulls in behind me at the railroad tracks. I have to stop here.
Anna:	You're at the stop sign before the tracks, in the village? Do you see anything there, or do you just drive through?
Beth:	There's lights there on the right.
Anna:	Do you see what's causing the lights?
Beth:	No, just blinking red lights or something. Maybe flares on the road.
Anna:	Okay—
Beth:	I'm not going to go look. I don't have to go that way.
Anna:	So what do you do now?
Beth:	Turning left...driving, faster.

Anna: Is the truck still behind you?

Beth: Um…. *[Anxiously:]* He can pass! There's nobody on the road! Be my luck to get a stupid drunk driver!

Anna: Is there anything unusual that you see around you?

Beth: *[Whispers:]* No! I just want to get away!

Anna: Do you know what time it is? Can you look at the clock?

Beth: Nine-thirty?

Anna: Okay. And what happens next?

Beth: It's the signal light, the tower thing, with the strobe light for aircraft. *[Impatiently:]* You know!

Anna: Uh-huh. Yeah.

Beth: I hate that light!

Anna: It's not so bright. It won't bother you.

Beth: I have to slow down.

Anna: Why do you have to slow down?

Beth: I have to slow down 'cause I have to cross the tracks. *[Anxiously:]* I can't go this fast!

Anna: Okay. Can you slow down?

Beth: *[Whispers:]* Yes…

Anna: Do you cross the tracks safely, or does—

Beth: Whew! Ho-o-o…. It's gone. Whew!

[At this point, I felt something may have happened—so did Anna! In an effort to find out, she asked me to backtrack to just before I saw the strobe light.]

Anna: Okay…. Let's go back to where you—

Beth: *[Sighs]*

Anna: Let's go back now…. Can you tell where he turned off? Did he turn off—

Beth: —don't look at it! It's right in front of your eyes and you have to look at it! I try not to look at it.

I was getting confused, so Anna tried to place me at the moment when I first noticed the truck was not behind me anymore. To me, this part had already passed and as I was actually reliving the incident, I couldn't go backwards because I had not done so then! Accepting that this strategy wasn't going to work, Anna allowed me to continue where I'd left off.

Beth: There's a lot of lights…. It's like an accident or some-thing. I don't know what that is…. It's a lot of lights. I can't—it's like I'm driving into a wall of lights!

Anna: Okay. Do you drive into that wall of lights?

Beth: I don't have any choice! I'm not going to go back and find the truck!

Anna: What happens when you get up to those lights?

Beth: It's like a roadblock or something…

Anna: Do you stop at the roadblock?

Beth: I have to…. He's waving me out, out on…the road. Gotta go that way, and…I can't go the other way! [*Surprised:*] Oh, geez, this is dumb!

Anna: What's dumb?

Beth: There's another roadblock to the right. I don't want to go out there!

Anna: Why don't you want to go?

Beth: I feel herded, herded! Like a cow!

Anna: Okay, so you stop at the roadblock?

Beth: Yes! Whew!

Anna: And then what happens?

Beth: There's men there…. Flashlights. God, this is dumb! I'm like five miles from home. Uh…inside light doesn't work. I need a flashlight, uh…. "Thank you."

Anna: Who are you talking to?

Beth: There's a man there.

Anna: And what does he look like?

Beth: Like a man.

Anna: Dressed in a uniform? Is he the sheriff?

Beth: No. The sheriff—there's a guy in a uniform…. Looks like the sheriff, out in front. He's got a brown leather jacket on, brown pants with a stripe down the legs.

Anna: The man that's talking to you, what does he ask you?

Beth: He wants my driver's license.

Anna: And do you give it to him?

Beth: [*Exasperated:*] Well, I have to find it first! [*Worriedly:*] The lights don't work and I don't want him to know! I need a flashlight. I don't want him to see I'm shaking!

Anna: Why are you shaking?

Beth:	Because I don't like this! I feel herded! [Frightened:] Because there's nobody else here! Why isn't there anyone else here? Why do they have to have *all these cars?* There's trucks, and over there there's jeeps, and camouflaged vehicles, police cars and unmarked cars—and there's all these cars! And they all have their lights on, and they're all, like, flood lights.
Anna:	Okay, do you find your driver's license?
Beth:	[Quietly:] Yes.
Anna:	And what does he say?
Beth:	"Thank you."
Anna:	Does he give you back your license?
Beth:	Yes…
Anna:	Does he tell you what the problem is?
Beth:	No…. He wants me to get out.
Anna:	And do you get out?
Beth:	[Excitedly:] I don't think so! I'm not getting out of the car! "Why?" Look, I'm just going home. You see, I go to college at night on Thursday nights. I would like to go home.
Anna:	And what does he say to that?
Beth:	"Please step out of the car." I can leave soon. I don't want to get out! I got my elbow on the door lock. [Relieved:] I got the door locked! Don't do that! Don't do—
Anna:	What happened? Tell me what happened.
Beth:	[Whispers excitedly:] He opened the door!
Anna:	Who opened the door?
Beth:	He opened the other door!
Anna:	Did someone else open the door?
Beth:	[Gasps] Yes! Don't do that! [Frightened:] You don't have to do that!
Anna:	And who is this person who opened the door on the other side?
Beth:	It's a man in a coat…
Anna:	And what does he say to you?
Beth:	"Step out of the car!" [Crying:] Why? What's the matter?
Anna:	Does he answer you?
Beth:	[Still crying] No-o-o-o! I haven't done anything wrong. [Angrily:] If you, if you want to stop somebody for some-

	thing, go get the crazy guy in the truck! I don't want to get out of the car! *[Gasps]*
Anna:	What's happening?
Beth:	We're outside the car, on the passenger's side.
Anna:	What's there, outside the car on the passenger's side?
Beth:	*[Breathing heavily:]* The man in the coat.
Anna:	The man in the coat. What does he do?
Beth:	Makes me get away from the car.
Anna:	Did you get out of the car?
Beth:	No. He pulled me out…the passenger's side. I didn't want to go.
Anna:	And what does he say to you?
Beth:	"Step away from the car, please."
Anna:	And do you?
Beth:	I don't think I'm going to argue with him now. There's a whole bunch of them here.
Anna:	What does he say to you?
Beth:	To come with him.
Anna:	And do you go with him?
Beth:	I have to! There's too many of them.
Anna:	Where do you go with him?
Beth:	To the truck.
Anna:	What truck?
Beth:	There's a, a panel truck with, like camouflage stuff on it—canvas on the sides. Like a troop truck…smaller.
Anna:	Do you get in the truck?
Beth:	They put me in the back of the truck. I don't get in.
Anna:	Does someone else get in with you?
Beth:	Yes.
Anna:	Is this someone in uniform?
Beth:	No. He's got a black coat on. *[Crying:]* What do they want me for? Tell me what they want me for?
Anna:	Did he ask you questions?
Beth:	*[Voice shaking:]* No-o-o…. *[Crying again]*
Anna:	What happened?
Beth:	Put a blindfold on me!
Anna:	Who put a blindfold on you?

Beth: Please! *What do you want? [Crying, voice shaking:]* Tell me what they want!

Anna: Do they tell you?

Beth: No! Where are we going—we're moving! They're taking me somewhere. Where are we going?

Anna: Do they answer your question?

Beth: No.... They have my arms, holding my arms down. I can't see anything.... We're going straight. *[Quietly:]* I'm going to pay attention. I need to pay attention! People can do this.... Going straight, straight...straight—

Anna: How long did you go straight?

Beth: Uh.... Think, think, think.... Uh.... Minute, minute and a half. Not too fast.

Anna: And did you turn off?

Beth: Yes.

Anna: To the right or to the left?

Beth: Turn right.... Straight again, straight. *[Breath quickens:]* It's hard to think! Oh!

Anna: It's okay. It'll be easy for you. You can do this. It'll be easy for you to do—

Beth: Now they're talking to me.

Anna: What are they saying to you?

Beth: *[Whispering:]* I think he's trying to keep me from thinking!

Anna: Keep you from thinking.... What does he ask you?

Beth: Questions. He's—

Anna: What questions?

Beth: "Where do I come from? Where do I live? Where was I born?"

Anna: Do you answer them?

Beth: No! I'm paying attention. They're distracting me! They're doing this on purpose. Oh, no!

Anna: What's happening?

Beth: Oh! Facing the other way...uh...they turned me around! Oh, shoot!

Anna: Is the car still moving?

Beth: Truck...yes, but.... Oh, shoot, I lost something! I don't know.... Okay. Okay...we're turning. *[Whispers:]* Oh,

	geez, okay…. Which way are we facing? *[Loudly:]* Don't talk to me!
Anna:	Are they still talking to you?
Beth:	They're going to move me again. *[Gasps]* I'm getting confused—
Anna:	It's alright. You can be confused. It's alright, it's not important.
Beth:	Yes, it is! I need to know where I am! I know this area, I can find things. I know these roads. *[Annoyed:]* They're confusing me!
Anna:	Okay. Let's go to the point—does the truck stop?
Beth:	*[Doesn't answer]*
Anna:	Do you get out of the truck?
Beth:	They carry me out.
Anna:	They carry you out. Where do they take you?
Beth:	I don't know! I don't know where I am.
Anna:	Are you taken into a building? Are you in woods?
Beth:	Inside, 'cause I can't hear anything, any noises. I can see bright—I can see light filtering through the blindfold. What? What?
Anna:	What did they ask you? Did somebody ask you something?
Beth:	—wants to know how long, how long have I known…things.
Anna:	What kind of things?
Beth:	*[Frustrated:]* I don't know! I don't know what he wants.
Anna:	What do you answer?
Beth:	"I don't know."
Anna:	Are you still blindfolded?
Beth:	Yeah. Yes. Makes me sit down.
Anna:	Are you sitting on a chair? On a couch?
Beth:	It's, huh…hard. Um…my hands are behind me…
Anna:	Are they tied behind you?
Beth:	Yeah, but not—not, it's like not the wrist, it's like the elbows are tied. There's a tie around the elbows, and then it's around my waist.
Anna:	Can you feel the back of the chair?
Beth:	It's a wall…

Anna: It's a wall? So you're sitting down, what happens next?

Beth: A man's voice. A—*[unintelligible]* Oh!

Anna: What happened?

Beth: *[Loudly, startled:]* He gave me a shot!

Anna: He gave you a shot? In your arm?

Beth: I think so…. *[Startled:]* Stuck me with something…

Anna: And how do you feel?

Beth: Mad! *[Yells angrily:]* What do you want?

Anna: And what do they answer you?

Beth: *[Quietly:]* They don't. *[Pause, then groans softly.]*

Anna: How are you feeling now?

Beth: *[Groggily:]* Sleepy…. I don't know…why…I don't know—oh-h-h. *[Pause]* It's so hard…to pay attention.

Anna: Yes, but you can do this very easily.

Beth: —moving it over there.

Anna: What over there?

Beth: I don't know…. I can't understand what they're saying. Can't open my eyes anyway so it doesn't matter.

Anna: Okay.

Beth: *[Pause, then in slurred voice:]* She'd just let me go to sleep…. They're going to move me again.

Anna: They're going to move you…. Where do you get moved to?

Beth: *[Whispers:]* I don't know…they just pick…me up. *[In wonder:]* It must be hard for them to walk backward.

Anna: Where do they take you? Where do you end up when they're carrying you?

Beth: I'm in…someplace, I'm, I, they have to take my clothes off.

Anna: Who takes your clothes off?

Beth: Help me take—

Anna: Are your arms still behind you?

Beth: —move everything around. *[Breathes deeply]*

Anna: Do they—your clothes have come off?

Beth: They just move them around.

Anna: What do you mean, they move your clothes around?

Beth: I don't know what they're doing!

Anna:	Anybody touching you or anything?
Beth:	Yes.
Anna:	Where are they touching you?
Beth:	I don't know…. Something warm, checking…. Huh! It's hard to concentrate. It's very hard to concentrate.
Anna:	It will be very easy for you to concentrate. Where does this something warm touch you?
Beth:	Uh, on the navel…. It doesn't touch. I can feel hands, navel…. *[Softly:]* Here *[touches abdomen]*…on the back…under my arms.
Anna:	Are they poking you, are they prodding you, or pushing?
Beth:	I don't care…. *[Dreamily:]* I don't…seem…to care.
Anna:	Is anybody talking while they're doing all this to you?
Beth:	*[No answer]*
Anna:	They're completely silent?
Beth:	Something…. There's a light bulb that makes noise. Sounds like that…
Anna:	Is there maybe a radio in the background?
Beth:	No. It's like…a fluorescent bulb…that's not, that's going to burn out.
Anna:	What happens after they finish touching you on your body? Are you back where you were?
Beth:	No…just move again.
Anna:	They pick you up again?
Beth:	No. Turn me…around, and…uh…I get to lie down. Maybe they're going to…let me…sleep now.
Anna:	How do you lie down?
Beth:	On my back.
Anna:	Your hands aren't behind you anymore?
Beth:	Yes…they're…behind me…and my arms.
Anna:	Are you on a bed with a nice soft mattress?
Beth:	No.
Anna:	What are you lying on, then?
Beth:	Feels like…a cot, or something. It's not real soft, but it's okay.
Anna:	And what happens while you're lying there? Do they start talking to you?

Beth: No.

Anna: Does everybody leave the room?

Beth: *[Annoyed:]* Let me sleep.

Anna: They're going to let you sleep?

Beth: *[Whispers:]* I wish you'd just go away. I don't want you to do that.

Anna: What don't you want—

Beth: *[Frightened:]* Spreading my legs…

Anna: Who's doing this?

Beth: A woman. It doesn't feel like a man. *[Worriedly:]* She's going to examine me there.

Anna: Does she use her fingers or does she use a speculum?

Beth: First, it's just that warm—it's probably a light. It feels warm…. She says, "Let's have a look." I'm trying to close my legs!

Anna: Can you close your legs?

Beth: *[Quietly:]* Almost…. Uh! *[Starts crying]*

Anna: What's happening?

Beth: *[Grabs right arm]*

Anna: You got another shot?

Beth: I think so…. *[Sniffles]*

Anna: *[After a long pause:]* Now what's happening?

Beth: Boy, am I thirsty!

Anna: Is somebody talking to you?

Beth: I can't hear anything.

Anna: Do you feel like you're alone?

Beth: I don't think so…

Anna: Did they give you something for your thirst?

Beth: *[Pleading:]* "Can I have something to drink, please?"

Anna: And what happens next?

Beth: Sounds. People moving around.

Anna: Do you hear any voices?

Beth: Whispering, but I can't tell—they sound like they're not very close to me. Sitting in a lounge chair, it's like a lounge chair with a high back on it. I'd like to get some blood moving in my arms. They hurt.

Anna: Do they respond to you?

Beth:	No, and I don't want to make them mad!
Anna:	Does anything happen to you while you're sitting in this lounge chair?
Beth:	I can smell something.... It smells like, like that stuff—what we used to use, like spray starch. Like they use it on their clothes. I can smell it and that's what it smells like.
Anna:	Okay, what happens next?
Beth:	The one standing next to me, he takes my arms from in front and lifts me up, and asks me if I can walk.
Anna:	Can you walk?
Beth:	*[Laughs]* I don't know! I haven't tried recently.
Anna:	Do you walk?
Beth:	Yeah, but I think I fall down.... I don't know where I'm going!
Anna:	Are you still blindfolded?
Beth:	Yeah. "Can I go home now?"
Anna:	Is anybody talking to you, are they asking you questions, are they telling you anything?
Beth:	No.
Anna:	And so you walk—
Beth:	*[Tearfully:]* "I just want to go home! I'll do what you want, just let me go home!" *[After a pause:]* Nobody's talking to me at all. I just want—leading me somewhere...
Anna:	Okay, when you get to that someplace else, where are you? Do you know?
Beth:	Went down a...slope...like a, um, like a handicap ramp.
Anna:	Are you inside or outside, do you think?
Beth:	I think I'm outside now.... It's colder, colder than it was.
Anna:	Is it dark outside, or is it the same amount of light you had before?
Beth:	It's still pretty light. I can see light...through the blindfold.
Anna:	And where do you go after the—
Beth:	*[Surprised:]* Whoops! *[Laughs]*
Anna:	What happened?
Beth:	*[Slurred:]* There's...a step here.... *[Hiccups]* "Well, thank you!"

Anna: What—

Beth: "Give me a hint!"

Anna: Do you walk down one step?

Beth: I hope! Up, stepping up…. One step…one step. A little warning would be nice! *[Quietly:]* I think I'm going to get to go…

Anna: And when you finish walking, where are you?

Beth: In the back of the truck, I think.

Anna: Is the truck moving again?

Beth: No, not yet….

Anna: Is somebody else in the back with you? *[Beth nods yes.]* More than one person?

Beth: Yeah. One on each side.

Anna: Can you touch them?

Beth: No. I mean, I can feel them—one here *[points to right]* and one here *[points to left]*. They're right up against my arms, so I can't move.

Anna: Does the truck start moving?

Beth: Huh! Here we go, here we go…. Oh, I just want to go home! …Moving, moving…. Musical chairs…. I'm trying, trying to concentrate. I can't concentrate!

Anna: That's okay—

Beth: I feel like I'm drunk! Off we go, musical chairs, musical chairs.

Anna: Okay. Now at some point the truck stops. What happens then? Do you get out of the truck?

Beth: Oh, oh boy! Ow-w-w…. *[Groans]* Okay….

Anna: That's okay, you're—

Beth: *[Grabs upper arms, sobbing:]* I was better off before!

Anna: Alright, your arms won't hurt. We're going to turn down that pain now. I'm going to touch your hand and turn down the pain. We're going to take it down from ten…to nine, getting less and less…to eight…to seven, much less now…to six—pain's going away, it's much less now…. Down to five…four—

Beth: Getting out of the truck….

Anna: Do you still have your blindfold on? *[yes]* Are you walking? *[yes]*

Beth: *[Dreamily:]* Here we go….

Anna:	You're going now?
Beth:	*[In a normal voice, but impatiently:]* They're moving the road flares out of the way so I can go. Finally! Geez!
Anna:	Are you standing beside the car?
Beth:	No! I'm in the car!
Anna:	They put you in the car before they took the blindfold off?
Beth:	*[Confused:]* What?

[I had no recollection of the previous events and couldn't understand why Anna would think I was outside the car or wearing a blindfold!]

Anna:	*[Long pause, also confused]* That's alright. You're sitting in the car, and—
Beth:	I'm in the car and I get to go now. Uh—
Anna:	You're still talking to the man that's standing near—
Beth:	He just waved me on…. I'm going out through the, the flares are gone. Whew, boy!
Anna:	Okay. Did you look at the clock? Do you know what time it is?
Beth:	11:15? That can't be right, can it? *[Pause]* I'm awful confused. I don't know what's wrong here….
Anna:	It's okay. We can be confused! What we're going to do is, we're going to leave this place. We know you made it home safely. Let's just relax right now.
Beth:	Is there something wrong here? What did we do wrong? *[Begins to wake up without prompting]*
Anna:	Don't worry about something being wrong. Just relax, be calm. You—
Beth:	*[Laughs]* I wasn't—I woke myself up, I think!

Both of us were startled by this turn of events, plus I was experiencing something that Anna was much more familiar with—having two separate memories of the same event! They were fighting for space in my head and I couldn't figure out which was fact and which was fiction!

I could still picture myself sitting there impatiently behind the wheel, watching the deputy move the flares out of the lane so I could pull through, totally unaware of anything having happened between the time I had refused to get out of the car until I was told to drive on! While still under hypnosis, only seconds after relating my return in the truck, I had completely forgotten the incident. When Anna asked me if I was standing beside the car, I had no idea what she meant! Why would I be outside the car? This confusion was probably what woke me up. Yet I *could* remember how my arms ached and how tired I was. When I saw the time, I was aware that something had happened—too much time had passed.

Another conundrum fascinated us: If this really had happened, this abduction by humans, why had they been so demonstrative? Wouldn't it have been simpler to just force me off the road while I was still in the boonies, where no one was likely to witness it?

And why so *many* people? Maybe I'd been selling myself short! Did they expect so much trouble from me that they felt the need for reinforcements? It didn't make any sense!

How did they manage to make me forget everything that took place while I was with them? Did we have the technology to induce selective memory loss? Was that what the injections were for?

There were so many questions, so much stuff that didn't seem to make any sense. I couldn't let myself believe this had really happened as I remembered it. There had to be another explanation.

What about a screen memory? I suggested this to Anna some time later. The gray shits have certainly been known to do that so that we don't notice the loss of time. She considered that for a moment, then countered, "Why bother? You've been getting conscious memories for a long time now." Then she added that if this had been a screen memory, it was inordinately complex and only served to make the abductee *more* curious, not less. Even so, I just couldn't accept what I remembered under hypnosis. I would need more proof.

Several months passed before I had an opportunity to reexamine this memory. A certified hypnotherapist, new to our area, had agreed to meet me for a consultation. Initially I had hoped to add her name to the roster of qualified therapists available to counsel abductees locally. If she seemed helpful, I would recommend her to the Fund for UFO Research and she could be referred to others in desperate need of help. I say "initially" because after our first meeting my priorities shifted slightly.

Marilyn Carlson was easy to talk to and an experienced hypnotist. She had helped clients who had reported being abducted, but did not "specialize." Many of her past and present clients were *not* abductees, and this immediately made me feel more comfortable around her. She knew about the phenomenon, but did not presume an abduction took place just because the client exhibited similar symptoms. It was her experience as a hypnotist, though, that attracted me. Maybe she could help me get to the bottom of this perplexing enigma.

We made arrangements for an hypnosis session the following week, intending to cover the roadblock incident from start to finish as if I knew nothing of what may have happened that night. Unfortunately, with very few exceptions, the same story emerged! How could it be real? I asked Marilyn afterwards.

"Well," she opined, "I don't know if this will help or not, but I can tell you I've heard of something like that happening to someone else. I'm not offering this as proof—I don't think you'll ever find that—but something very like what you've told me was reported by one of my other clients. I don't understand it, I can't give you any reasons why the authorities would do this and then deny any knowledge of it, but it's how you remember it."

"But I don't think I'll ever be able to accept this memory as real," I insisted.

She smiled. "Maybe not, but that's okay. You don't have to accept it. Maybe now you can at least just put it away and try to forget about it. You got through it fine. You'll be okay. I think you're coping pretty well, considering."

I guess I was coping pretty well. Being angry sort of cleans out the pipes.

It made me think, though: I had previously believed that the government knew more than we ever would about this phenomenon and these abductions of people, but now I had second thoughts. If they knew it all, they would have no need to question us, no need to spy on us or tap our phones. So they know we're not alone...that certainly was not news to most of us, but perhaps they don't know how we are affected by these abductions, both physically and emotionally. If the local authorities came to my door and asked me to tell them what the aliens had been doing with me, I don't think I'd be inclined to confide in them! No doubt they know this, too.

For a long time I was angry over this confusing memory—not only because I could neither prove it nor disprove it, but because I was afraid. I was afraid it was true, and if it was, would it happen again? It was bad enough being abducted by aliens; how could we defend ourselves against abduction by our own species?

But fear debilitates; it makes us hide inside ourselves. We can't act when afraid—we can only *react*. And it feeds on itself, becoming stronger and more dominant the longer it is in control. I couldn't live in fear, and I didn't want to be angry, so I filed the experience away. Maybe one day I would understand it.

69. Anna

Beth's recall of her "military incident" was more detailed than I expected, but not particularly surprising. I'd read many stories by others who felt that some aspect of military intelligence knew much more than was being revealed publicly. We'd not been approached openly by the military, nor had we been harassed by the alleged Men In Black, as so many other abductees reported. Yes, we'd had our phones tapped and been followed around by men in dark sedans, but I needed something more to convince me that my own government was in league with the aliens. Beth's abduction by humans was it. Our government is up to it's epaulets in this. They know.

The military cover-up theory surfaced during one of my sessions with Norris Blanks. I trusted Norris completely to see that he did no harm to me with our therapy, but he sometimes led me in directions that revealed where *his* concerns lay, not necessarily my own. He allowed me to bring out memories that I hadn't even been curious about. We were again talking about the implants in my body, and he was asking my subconscious to reveal what it knew about them.

Thursday morning, November 18, 1993

Norris: I want you to ask about the three black squares. I want you to ask if it is appropriate that they be there.

[Note: There were only two black squares, implants—the one in the back of my head was a thought implant and did not show up as black.]

Anna: Yes.

Norris: Ask what they represent and what they mean.

Anna: For my protection.

Norris: Protection from what?

Anna: Interference.

Norris: Interference from who?

Anna: *[Garbled response]*

Norris: Which government?

Anna: U.S.

Norris: Ask why the U.S. government would want to interfere with you.

Anna:	*[Garbled]*
Norris:	Ask when these plates, can we call them plates? Plates or do you have a better definition of them? What would you name them if you had a definition of them?
Anna:	Plates is okay.
Norris:	How long have these plates been there?
Anna:	Many years.
Norris:	When did the first one come in?
Anna:	Three years.
Norris:	You were three years old?
Anna:	Yes.
Norris:	Which one was that?
Anna:	The head.
Norris:	I want you to ask now what would happen if you removed the implant in the back of the neck.
Anna:	They'll just give it to someone else.
Norris:	What harm could come to you if the implant were removed?
Anna:	Confusion.
Norris:	Ask to be told what the U.S. government is doing to some of its citizens. Which part of the government is initiating this?
Anna:	Aurora.
Norris:	Ask if Aurora is motivated by the dark energies.
Anna:	No.
Norris:	Do you know what Aurora is?
Anna:	It's a secret.
Norris:	Where is Aurora based?
Anna:	A lot of places; it's all over the place.
Norris:	Over the U.S.?
Anna:	Over the world.
Norris:	Did it originate in the U.S.?
Anna:	No.
Norris:	Where did Aurora originate?
Anna:	In France.
Norris:	In what year?
Anna:	1947.

Norris:	And who were the people behind Aurora?
Anna:	Secret.
Norris:	Were they Earth people or alien people?
Anna:	Earth.
Norris:	Were they people who were being manipulated by the darker energies?
Anna:	No.
Norris:	Were they financial people? What type of people?
Anna:	Military.
Norris:	Are they in league with alien technology? Are they using alien technology?
Anna:	When they can.
Norris:	Is the core of this group European or American?
Anna:	American.
Norris:	Does the President know of this, President Clinton?
Anna:	No.
Norris:	No. Do people in the U.S. government generally know of this?
Anna:	No.
Norris:	Do people in the military know of this?
Anna:	No.
Norris:	Did George Bush know of this?
Anna:	Yes.
Norris:	Yes. What is their plan?
Anna:	To avert destruction.
Norris:	So what are they doing to people? How are they affecting people?
Anna:	They're taking care of the changelings.
Norris:	Ah, tell me about the changelings.
Anna:	*[No response]*
Norris:	That's okay, you can see this easily. You need not think about it and you are in a very protected place. That's it, take in a breath. Ask now about the changelings.
Anna:	It's…. *[Garbled]*
Norris:	Yes.
Anna:	*[Garbled]*
Norris:	When were you altered? At birth or after birth?

Anna:	Before.
Norris:	Before birth. And what is your mission here on earth?
Anna:	To change.
Norris:	And who do you owe your allegiance to?
Anna:	Myself.
Norris:	Right. And so the Aurora group, they are monitoring you changelings?
Anna:	Yes.
Norris:	How many changelings are here on the earth? Approximately.
Anna:	8 million.
Norris:	Let us talk further of the changelings. When did the changelings first get—
Anna:	A long, long time ago.
Norris:	Hundreds of years ago, thousands of years ago?
Anna:	Thousands.
Norris:	Are the changelings all associated with the extraterrestrials?
Anna:	Yes.
Norris:	When did the Aurora people first become aware of the changelings?
Anna:	1947, they found out. There was a mistake made.
Norris:	What do they think of this Aurora group?
Anna:	I guess they feel that they have to cooperate with them. I don't know why.
Norris:	How are they cooperating with them?
Anna:	Non-interference.
Norris:	Well, do they respect the rights of non-interference?
Anna:	No.
Norris:	Do they believe that there is a force in the universe that is above them, a higher force?
Anna:	No.
Norris:	They believe that they are the highest force in this universe.
Anna:	No, there's some other guys up there. Different guys.
Norris:	And they are above the ETs?
Anna:	Maybe. Probably.

Norris:	Why don't they just come down here and alter the course of human history? They obviously have the power.
Anna:	It's too big; it's natural stuff.
Norris:	Why don't they come down here and inform the world leaders?
Anna:	No one would believe them.
Norris:	I think many people would believe them. Many more people believe in these things now.
Anna:	That's what they tried when Aurora started.
Norris:	How did they try?
Anna:	Tried to tell people what was going to happen. Some people believed them, but....
Norris:	I call upon the one, now, that is giving the information, beyond the Anna entity. Will that one come forward and speak with me? Yes or no?
Anna:	Yes.
Norris:	Thank you for coming forward and speaking with me. May I inquire as to your name?
Anna:	Sonna.

[Note: I really felt his name was Sontag. I just couldn't pronounce it then.]

Norris:	Sonna, and are you the gray being that accompanies Anna?
Anna:	No.
Norris:	Where do you fit into Anna's spectrum of beingness?
Anna:	I visit sometimes.

[Note: My voice *deepens* here and takes on a different character on the tape.]

Norris:	And are you associated with the gray ones?
Anna:	Yes.
Norris:	Are you of their kind?
Anna:	Yes.
Norris:	Do you understand the reason for my questions?
Anna:	No.
Norris:	I speak to you to gain knowledge so that I may be able to help the Anna entity.
Anna:	She doesn't need help.

Norris:	But she isn't happy.
Anna:	Doesn't matter.
Norris:	Well, happiness is of concern here. Do you not care about her happiness?
Anna:	No.
Norris:	Is she just an object to you?
Anna:	A very valuable one.
Norris:	A valuable one, right. Is there one that stands behind you, Sonna? A dark one?
Anna:	No.
Norris:	If I call upon the Legions of Light, Archangel Michael, to bring forth the nets of light to comb Anna's aura, would this be acceptable to you?
Anna:	It won't matter.
Norris:	It won't matter. It will have no effect?
Anna:	None.
Norris:	How long have you known Anna, Sonna?
Anna:	Forever.
Norris:	How long is forever?
Anna:	Thousands of years.
Norris:	Thousands of years. Did you ever have an Earth body yourself?
Anna:	No.
Norris:	Have you always been one of the ET kind?
Anna:	Yes.
Norris:	Yes, so you attached to Anna thousands of years ago? What is the plan here, Sonna? Are your people looking to use the Earth race to create another civilization?
Anna:	To populate planets.

[Note: My voice has *deepened* further and slowed in speech patterns.]

Norris:	I see. So you are looking for ongoing colonization, are you?
Anna:	Yes.
Norris:	Were you the original colonizers of the planet Earth?
Anna:	Yes.
Norris:	Is there any way that Earth people can be assured of that?

Anna:	They'll never know.
Norris:	So, in a sense the Earth people are like puppets. Is that true?
Anna:	They're just one of the species on Earth.
Norris:	Was your kind the originator of most of the species?
Anna:	No.
Norris:	Is there one above you who you take orders from?
Anna:	No.
Norris:	May we speak again sometime later, Sonna?
Anna:	Yes.
Norris:	I thank you for conversing with me and enlightening me as to this process. Will the Anna being now come back? Did you hear that, Anna?
Anna:	Yes.

[Note: My voice *changed back* to normal.]

Norris:	Good. And how are you feeling?
Anna:	Okay.
Norris:	It's okay, because you remember that we are all part of an adventure here.
Anna:	A *weird* adventure, but yeah.
Norris:	It's all part of an adventure. Just like you're told you'll go, Norris is told he'll stay, and others are told other things.
Anna:	But they lie.
Norris:	Yes, I agree with you. Do you feel much of what we just heard is lies?
Anna:	No.
Norris:	No.
Anna:	*[Garbled]*…talking—there's time. It's not that long to wait. 1997 isn't that far away. Things have already started to change.

End Transcript

I didn't know what to make of this. Was it confabulation? I knew that the first cover-up occurred in 1947. A flying saucer crashed near Roswell, New Mexico, in July of 1947. The military issued a press release to that effect, and then quickly changed their story when the top brass got involved. They said it was a weather balloon. Not too many people are gullible enough to believe that a crashed weather balloon can leave debris scattered over miles of range land, or that it would require such intense military secrecy all these years. There's too much evidence to believe it didn't happen. But my subconscious said the cover-up occurred in France, not New Mexico. France has had its share of sightings over the years, and continues to have abductions and sightings to this day.

Since that session, I have learned that Aurora is the name of a supposedly top secret, black project airplane being developed by our military. I was not aware, consciously, that George Bush had been Director of the CIA. When Bob Huff told me this, it made sense that if anyone in the government would have knowledge of a military cover-up, it would be him. I think it was incidental that as President he had this knowledge. I feel that U.S. Presidents are not routinely told about the military's involvement with aliens, nor about their abductions of citizens.

I didn't quite know what to make of Sonna. I seemed to take on a completely different personality, mannerisms and tone of voice when he was speaking. Was it really an alien talking through me, or just a new personality from my subconscious? I don't know if either explanation is possible.

I still take Norris's advice and treat most of this as an adventure, but it is definitely a weird adventure.

70. Conclusion

Opening up to others was easier than we had ever expected. No longer so concerned about how we might be perceived by family and friends, the "horrible secret" lost its power over us. This release provided numerous benefits: We felt happier than we'd felt in months; we looked forward to a future that still held promise; and we found new friends, people we might never have had the opportunity to meet otherwise.

Anna

I looked forward to meeting with other experiencers, but I was also a little leery. What if they turned out to be weird? I knew the people I had met at Budd's support group weren't. Beth and I surely weren't, but I didn't know about these other people. The first meeting dispelled those fears; we were all pretty much normal. Of course we were each dealing with our abduction memories in different ways, some better than others at this time. We were no different than any other group of people who meet because of shared interests. The only real differences I noticed were age and occupation. I had not been with such a diverse group in a long time. Our ages ranged from young children, through teenagers, to adults well into their sixties. Our occupations were equally diverse; professional people, blue-collar workers, homemakers and the unemployed. Most were experiencing current abductions, but those who weren't openly admitted to maybe not remembering.

I must admit to being a little distanced from the group for the first few meetings. I still found it hard to be around the children, so I concentrated on talking with adults whose children were not present. I also had another agenda in mind: I wanted to find out what we had in common. There must have been something we all shared, something that made the gray ones want us. I needed to find out what it was.

I had already tried to find some connection—besides abductions—between Beth and me. We didn't share the same blood group, didn't have the same length toes or any other similar physiology, we had different ethnic and religious backgrounds, had completely different personality types, and different tastes in food and clothes. In fact, it was hard to find anything we had in common, besides our passion for horses and our admiration for one another. I didn't have any more success with this agenda in talking with other experiencers, and soon gave up.

I did find several other people who felt that the aliens had united them with their life partners, especially husbands and wives. Others were not so lucky; their life partners were married to someone else.

I soon gave up analyzing my new found friends and started enjoying their company. It was refreshing to be in a group of people where I could talk freely, where I didn't have to lie about the impacts of the gray ones on my life. Although we didn't dwell on it, many of us talked about our experiences, offering support, encouragement and ways of coping with current crises. But most of all it was a party.

I felt myself coming out of my self-imposed isolation. It felt good to laugh again, to poke fun at the gray ones, to enjoy being alive and to be with people I could care about.

Beth

At first, I couldn't see how getting together with other experiencers to hash over terrors could possibly accomplish anything. We didn't need more discussion, more expressions of helplessness. What we needed was a release!

But I was graciously informed that this support group was not being organized so that we could relate our individual horror stories; it was being formed so that experiencers could get together and socialize—just like regular people. Like a block party? Yes, like a block party. Then it sounded good to me.

Our first meeting, though, did not get off to the best possible start. Being inexperienced with this kind of get-together (and with these kinds of people), we had all agreed that some ground rules would have to be set, so that we'd all have a chance to express our opinions and offer suggestions about where to meet, how often to meet, and how best to assure a party atmosphere. So we began by making it a formal meeting! A hotel conference room was selected and each of us paid a share of the cost for this space. The room itself, which was too small for our gathering, had no windows and felt claustrophobic. There was a shortage of seating and many of us found it necessary to either cram our chairs on top of each other or stand up against the wall.

Facing our counterparts across a glass-topped conference table, many of whom we had never met before, did not inspire camaraderie or social interaction. But some things were accomplished in this stiff and unnatural setting: One, that dwelling on our abduction ordeals would be counterproductive; two, that gatherings should take place whenever most of us could attend; three, that entire families were welcome, whether or not the children were involved; and four, that we would never again turn our party into a business meeting!

Since that hesitant and naive beginning, our local support group (now totaling over thirty men, women and children), have gotten together whenever possible, usually on a weekend evening in one of our homes. Each of us contributes food and a positive attitude, allowing ourselves the luxury of an evening out that does not include the gray shits! Though tales of recent abduction experiences do sometimes come up in conversation, they are rare and seldom darken anyone's mood. This *is* a support group, after all, and if one of us needs a listening ear there is always someone willing to provide it.

The most satisfying aspect of these "parties" is the release of tension through humor. Here we can laugh at ourselves, joke about the aliens, speculate wildly over the phenomenon, and generally have a grand time at the gray shits' expense.

The abductee's spirit is not dead; it is only napping and just needs an occasional kick in the pants.

Epilogue

Anna

Are we really being visited by aliens from another planet, or another dimension? I think so; it's real for me. I can no longer hide in the delusion of insanity. I've experienced too much. I've learned too much.

I feel that I have been given an unpleasant task to do: Tell of the impending doom. I don't do it. I wish I could have been given a more amiable message, one of hope, and peace, and love. But that is not to be. When talking with others at work, I find myself curbing the instant mental response, "It doesn't matter." Even if I feel, some days, that there is no use in planning for the future, I have no right to frighten other people by inflicting them with my paranoia. I know the aliens tell lies and use psychological testing to elicit emotional responses, but sometimes I find it hard not to believe them.

Or maybe we are just part of the conditioning process. The casual acceptance of aliens is widespread in the U.S. and abroad. More people than ever believe in the *possibility* of aliens visiting our planet. It's the actuality of it that will be the leap of faith. As more people regain their memories, as they begin to talk more openly about what they have experienced, acceptance will become more reasonable. We are connected to the Earth and to the aliens; what better intermediaries? That assumes some purpose for their revealing themselves. I don't know what that might be. It certainly doesn't seem to be to save us from our own destructive habits.

Their genetic agenda is confusing to me. Everything I read implies that they are developing hybrids. In my own case, I have never knowingly had sex with an alien, although I don't know what types of embryos they may have implanted in my body. I'm convinced that my alien children are genetically fully human, but they have been developed in the laboratories, outside a uterus, and no longer look completely human. If, as they have told me, they are developing beings to colonize other planets, then why do they need me? I'm especially suited to survive on planet Earth. Are there other planets so similar to Earth, and uninhabited, where we can survive? Are we so unique that it is worth all of the time and trouble? Why not develop clones that do not require the traumatic collection of sperm and ova? It seems there are always more questions, few answers. I keep looking for patterns, but if they exist, they are obscure. We need more resources devoted to these mysteries. It's time to take this seriously.

I no longer have the need to control the environment or the people around me. I am more accepting of what is, for it's own sake. This assimilation was not easy. The grays give me no joy. They stole my childhood by devastating my family relationships, by taking away my memories and supplanting them with others. I hate them for allowing me to believe that my father raped me. How many other children have they done this to? I'm convinced that some of the people who have recalled childhood abuse, especially when the reported incident was never repeated and the accused parent adamantly denies the abuse, have been given screen memories to cover the trauma of alien abuse, not parental abuse. Working with open-minded therapists has finally allowed me to accept what really happened to me. I have come to know that relationships with other people are the most important things in our lives. Everything else is just a way to structure time. Little else really matters.

The past few years have been an awakening experience for me. Not just dealing with the alien interference, although that would have been enough. The process of uncovering hidden memories has forced me to examine the very heart of myself. I'd never have done it willingly; I wasn't that type of person. It's been a journey into the deepest part of myself, a laying bare of who I am, what I think I am, and what is really important to me. I look at each new season, each new vista of this planet I call home, each person I interact with in a different light. Part of me will always question, "Will this be my last spring (sunset, Christmas, lover or horseback ride)?" Another part of me just enjoys the sensations, savoring each new experience as a reflection of truth, not necessarily truth itself.

No matter which interpretation of these strange events in my life is shown to be real, I have learned to accept myself. I like the new me.

Beth

Why me? I've asked myself that question countless times. And the answer is always the same: I don't know. I have entertained a number of theories, though. The most credible seems to be that these beings have been abducting generations of families, and may still be doing so. Whether this pattern would stand up to close scrutiny is anybody's guess. By the time I started getting memories, my grandparents were both deceased, so there was no way to verify the odd stories related about them over the years.

My father, fortunately (if that's the right word here), remembered a great deal about his childhood experiences and had documented many of them, but what if he had refused to talk about them? He could just as easily have reacted to my inquiries as my mother had, by evading them or leaving the room whenever the subject arose. And what if I hadn't had the courage to confront him? I don't imagine he ever would have broached the subject with me!

There were many strange events during my son's early years that I had conveniently written off as coincidental or just unexplained—until I actually verbalized them. Even after I had memories of being abducted as a small child, I did not want to associate the similarities between my son's night terrors and my own with this phenomenon, because by doing so I could be exposing myself to ridicule. Despite what I said aloud, I really didn't want my son to think I was crazy! When my granddaughter presented her drawing to me, describing these beings and their craft in such vivid detail, I didn't want to accept that either, but in her case my reasons for denial had little to do with my image. I simply couldn't believe these abductions have been going on for four generations! It was easier to accept the notion that our family tree harbored some unidentified genetic defect.

Then other reports surfaced of multiple generations having been abducted. Did this mean that this business had been going on prior to the infamous 1947 purported crash landing of a flying disk? All along I had thought that was when it had started—that flying saucers were unheard of before the early forties, and that abductions were unknown until the Betty and Barney Hill incident in the sixties. Considering the time period, it would be too much to expect that anyone in his right mind would *voluntarily* report being abducted by aliens! Even now, to stick one's neck out by publicly admitting involvement is to chance decapitation.

The populace, as a whole, *is* more accepting than it once was, and this new attitude is evident in almost every aspect of our daily lives. Turn on the TV, and what do you see? Hamburger commercials using a logo that suspiciously resembles a UFO, both in appearance and movement; another fast-food establishment uses commercials with varying scenarios involving the landing of UFOs outside a restaurant while patrons go on eating, presumably because the food is so good they cannot be distracted by such a common sight! Weekly programs highlighting sightings and alien abduction reports are more common—and more popular—than ever before. Is this a hint? Or is it only conjecture? Are we being *programmed* for acceptance? If so, by whom and for what purpose?

I have suffered through a roller-coaster ride of emotions in an effort to come to grips with this weirdness. I have denied it outright, accepted it fully, and wavered between the two, but at no time did I consider ignoring it. There have been entire days when I didn't think about it at all, not because I had learned to assimilate—or compartmentalize—but out of pure stubbornness! I wanted to feel in control of my life again; I wanted to know that my actions were the result of personal choice, not alien intervention. Completion of the simplest tasks was an enormous boost to my self-esteem.

The will to survive—and prosper—despite these intrusions, has not been supplanted by the need to reach some conclusion. I have questions, of course—lots of them! I have doubts, too, and suspicions and worries and theories. But what I don't have is my old, safe life. That, clearly, is gone forever. It had only been a veneer after all, so the process of discovering what now lies exposed should be a challenge. Am I up to it? Is anyone? We know more than we did before, but not enough. And there the real challenge lies: Will we recognize and accept understanding when and if it comes?

Again, we are plagued by questions and not enough answers, and speculations do not adequately address the emotional consequences. So many lives have been—and are now—affected by this controversial phenomenon. We can no longer afford the luxury of ignorance. Something is certainly going on here, and as both Anna and I have expressed throughout this book, it is past time for serious examination by the scientific community. We have been told, on a number of occasions, that the abduction phenomenon has proven difficult to explore under scientific guidelines. Why? Because, according to the experts, there is nothing concrete to examine, no physical evidence that would stand up under close inspection.

It's true that this phenomenon, unlike other mysteries, has presented scientists with little in the way of hard evidence, and that the testimony of witnesses also cannot be confirmed (except by other witnesses), but that does not mean it isn't worthy of serious consideration. Human memory is not flawless, but it certainly can be credible! Repressed memories retrieved through hypnosis can be confabulation, but how does that discredit conscious memories evident without the use of hypnosis? It doesn't. And in turn, it also doesn't prove that retrieved memories must be either confabulation or the result of leading by the hypnotist.

It is this human factor that so flusters the scientists. We are not *provable!* We are only human. Maybe it is this very aspect that has so fascinated our visitors.

We *are* unique, yes, but not because we are alone in the universe; we are unique because we crave understanding and knowledge, plan for our future, and embrace the lessons of the past. Once the earth was thought to be the center of the universe; once the world was believed to be flat; once humankind considered space travel impossible. If we have learned anything from our past philosophies, we should have learned that nothing is impossible, nothing is permanent, and truth lies somewhere between interpretation and understanding.

Anna Jamerson and Beth Collings.
(Photo by Daniel Light)

Author Profiles

Beth Collings is a professional horse trainer, riding instructor, lecturer/consultant and farm manager for a breeding and training facility in central Virginia. She has taught riding lessons, organized seminars, and handled promotion and advertising for the facility since 1987. She has previously held positions as riding instructor, supervisor for group life insurance for two major American and Canadian insurance companies, recreational counselor for the emotionally and physically challenged, police artist and free-lance writer for national horse magazines. Born in Washington, D. C. in 1946, Beth has lived in Virginia most of her life. She is divorced and has a married son and grandchild.

Anna Jamerson has owned and operated a horse breeding and training facility for over ten years. In conjunction with this activity, she has also developed and presented many seminars, taught riding, and lectured at local horse clubs. Since 1971, Anna has dedicated her career to the conservation of our beautiful and pristine wild lands. That year, she was hired by the National Park Service and worked everywhere from the Grand Canyon to Yosemite to the Virgin Islands! In 1975, she took a job with the U. S. Forest Service, and has been there ever since. Born in 1949, Anna is single and now lives among the beautiful rolling hills of Virginia's countryside.

Both Beth and Anna have attended many UFO conferences, as well as spoken at them. They continue to participate in panel discussions and give lectures. They have also appeared on numerous radio and television shows, speaking out about their experiences.

Swan•Raven & Co.

TAROT OF THE SOUL
A guiding oracle that uses ordinary playing cards
Belinda Atkinson

PLANT SPIRIT MEDICINE
Healing with the Power of Plants
Eliot Cowan

CALLING THE CIRCLE
The First and Future Culture
Christina Baldwin

RITUAL
Power, Healing and Community
Malidoma Somé

HUMAN ROBOTS & HOLY MECHANICS
Reclaiming Our Souls in a Machine World
David Kyle

WHEN SLEEPING BEAUTY WAKES UP
*A Woman's Tale of Healing the Immune System and
Awakening the Feminine*
Patt Lind-Kyle

For a complete catalog of Wild Flower Press or Swan•Raven books
or information on additional books that we distribute,
call **800/366-0264** or write to

Blue Water Publishing
PO Box 726
Newberg, OR 97132

Internet URL: http://www.bluewaterp.com/~bcrissey
e-mail address: BlueWaterP@aol.com

An audiotape and videotape recording of Beth discussing her abduction experiences is available.

The 90-minute recording was made in 1994 at

"The UFO Experience" Annual Conference.

For more information, write to
Omega Communications
P. O. Box 2051
Cheshire, CT 06410-5051, U.S.A.